NORTHERN PLAINS PUBLIC LIBRARY
3 3455 00726 5470

D0793193

The Big Bang

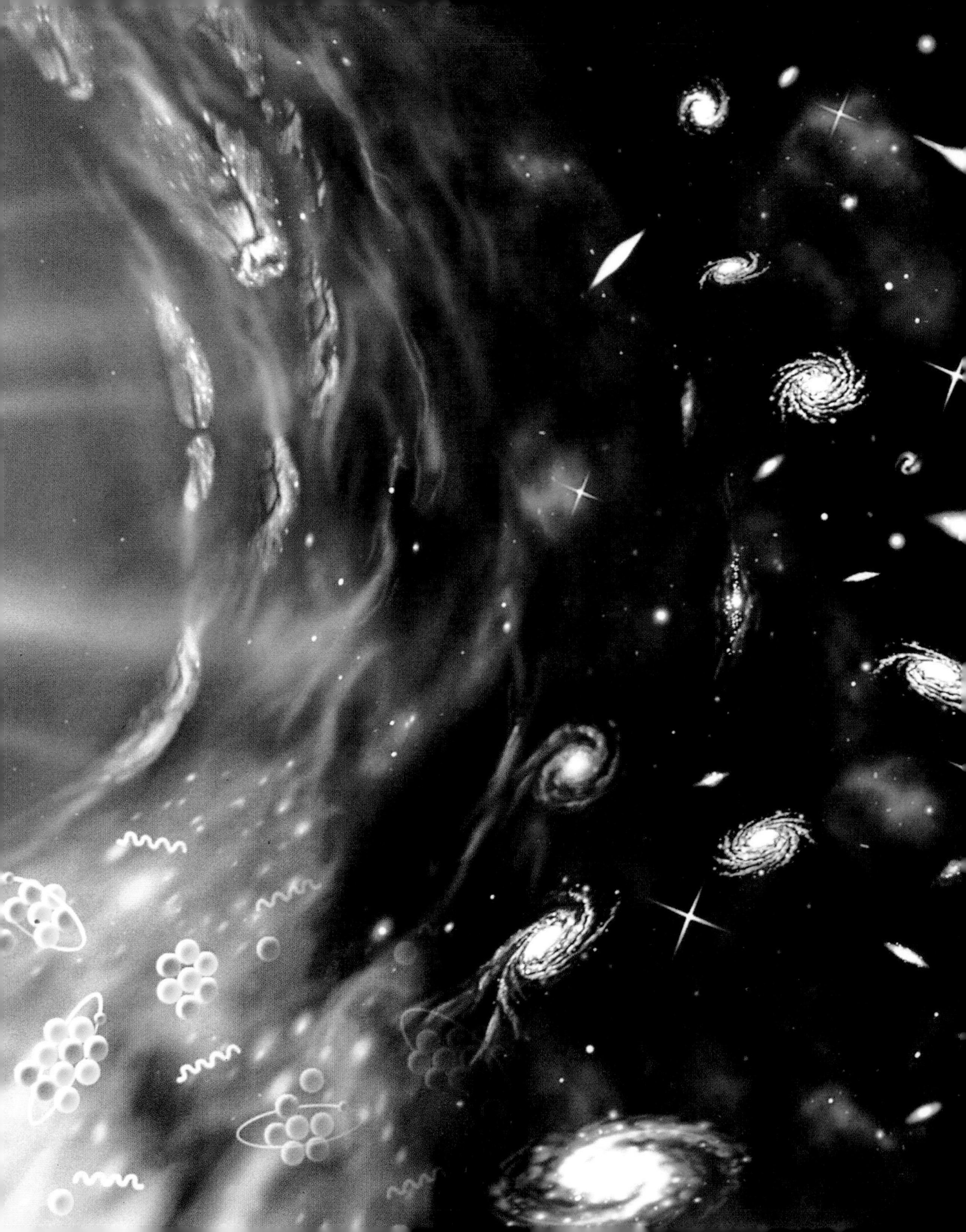

The Big Bang

Paul Parsons

First published in 2001 by
BBC Worldwide Ltd,
Woodlands, 80 Wood Lane,
London W12 0TT

DK PUBLISHING, INC.
www.dk.com

Publisher: Sean Moore
Art Director: Dirk Kaufman
Editorial Director: Chuck Wills

First American Edition, 2001

00 01 02 03 04 05 10 9 8 7 6 5 4 3 2 1

Published in the United States by
DK Publishing, Inc.
95 Madison Avenue
New York, New York 10016

ISBN 0-7894-8161-8

Produced for BBC Worldwide by
Toucan Books Ltd, London

Cover photograph: Mark Garlick/Science Photo Library

Printed and bound in France by
Imprimerie Pollina s.a.
Color separation by Imprimerie Pollina s.a.

PICTURE CREDITS:
Page 2 SPL/Henning Dalhoff/Bonnier Publications 6 The Art Archive/British Museum/Eileen Tweedy. 9 Hulton Getty. 10 Ann Ronan Picture Library, TR; The Art Archive/ Galleria degli Uffizi/Daglia Orti, BR. 11 The Bridgeman Art Library/ Louvre, Paris, France. 12-13 SPL/ T & H Hallas. 14 The Art Archive/ British Library. 15 The Bridgeman Art Library/Private Collection. 16-17 Anglo-Australian Observatory/Photograph by David Malin. 18 Hulton Getty. 19 Galaxy Picture Library, TL; Julian Baker, BR. 20 Hale Observatories, courtesy AIP Emilio Segre Visual Archives, TL; Galaxy Picture Library, BR. 21 The Hale Observatories, courtesy AIP Emilio Segre Visual Archives, TL; SPL/Peter Bassett, TR. 22 AIP Emilio Segre Visual Archives, Shapley Collection, TL; AIP Emilio Segre Archives, TR. 23 SPL/David Parker, TL; Julian Baker, BR. 24 The Bridgeman Art Library/Académie des Sciences, Paris, France, TR; Julian Baker, BR. 25 Galaxy Picture Library. 26 Corbis/Bettmann. 27 Corbis/ Hulton-Deutsch Collection, TL; SPL/A. Barrington Brown, TR. 28 SPL/Patrice Loiez, CERN. 31 Galaxy Picture Library/STScI/ NASA. 33 SPL/Dr G. Efstathiou. 34 AIP Emilio Segre Visual Archives, Lande Collection. 35 SPL/Mehau Kulyk. 36 Julian Baker. 37 Julian Baker. 38 SPL/ Sally Bensusen. 39 SPL/CERN. 40 SPL/David Parker. 41 SPL/ Mullard Radio Astronomy Laboratory. 42-3 Julian Baker. 44-5 STScI/NASA. 46-7 SPL/ Ducros/Jerrican. 48 SPL/Mark Garlick. 49 Julian Baker. 50 SPL/ Yannick Mellier. 53 Corbis/ Bettmann. 54 SPL/Frank Zullo. 55 Corbis/Roger Ressmeyer. 56 SPL/NASA, TR; SPL/NASA, BL. 57 SPL/CERN. 58 SPL/Ken Eward. 59 SPL/Celestial Image Co. 60 Corbis. 61 Corbis. 62 AIP Emilio Segre Visual Archives, Robin Collection. 63 SPL/Luke Dodd. 64-5 Corbis. 66 SPL/ NASA. 67 SPL/David Parker. 68 SPL/Lynette Cook. 69 Corbis/ Reuters NewMedia Inc. 70 SPL/ STScI/NASA. 71 Julian Baker. 72 SPL/Julian Baum. 75 SPL/ Mehau Kulyk. 76 Photograph by Richard Erens, courtesy AIP Emilio Segre Visual Archives. 77 SPL/Chris Butlere. 78-9 STScI/AURA/NASA. 80 DRK Photo/NASA OSF. 81 SPL/Julian Baum. 82-3 SPL/ Julian Baum. 84 SPL/Chris Butler. 85 Nick Ball/Mark Hindmarsh/Graham Vincent/ Sussex University. 86 SPL/ Lynette Cook. 87 SPL/David Parker. 88 SPL/Tony Craddock. 89 SPL/Julian Baum. 90 SPL/ David Parker. 91 SPL/Roger Harris. 92 Professor Sir Martin Rees. 93 SPL/Debby Besford.

Contents

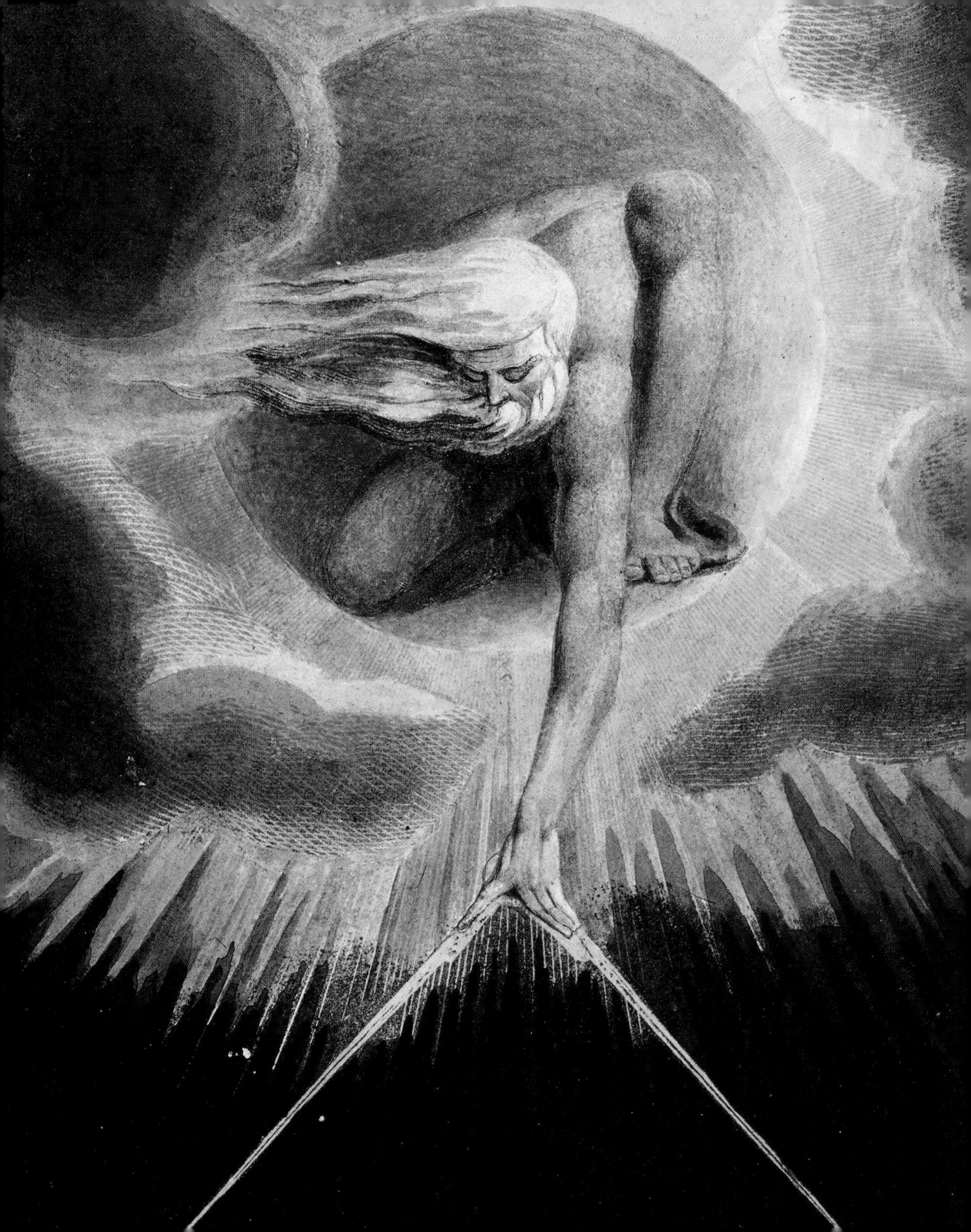

1

IN THE BEGINNING

1 IN THE BEGINNING

Suppose you had the patience to flip back the pages of the cosmic calendar one day at a time. After turning about 5.5 million million pages, going back some 15 billion years, you would finally come to a very special day indeed: the day our Universe was born. On that day – without warning and for no special reason – matter, radiation, space and time suddenly burst into existence in a searing hot fireball: the Big Bang. As the fireball expanded and cooled, galaxies and stars condensed out. Around some of these stars, planets formed. And on at least one planet chemical processes gave rise to what we call life. But this picture of our cosmic origins hasn't always been so clear. The ancient Chinese believed that the Universe was formed when a primordial cloud separated out into the heavens and the Earth. Other civilizations thought that the cosmos was made from the body parts of gods or mythical creatures.

Previous page: The Universe began some 15 billion years ago. Understanding that moment of creation has taxed scientists and philosophers for centuries. This painting, *The Ancient of Days*, is by William Blake (1757–1827).

THE BIRTH OF COSMOLOGY

Early ideas about cosmology – the study of the Universe – took their cue from religion. The Chinese in the 1st century BC believed that our Universe began life as a shapeless, abstract cloud. Those elements of the cloud that were pure and clear rose up to become heaven. Those that were heavy and opaque sank down to become the mortal world: the Earth. The ancient Chinese believed that pure, heavenly materials merged together more easily than those of the less pure mortal world, and so it was that the heavens formed long before the Earth.

Even further back, in the 12th century BC, the Mesopotamians believed that the god Marduk cleaved the body of the primeval mother, Tiamat, in two – one half formed the Earth, the other the heavens. On the other hand, Polynesian myths tell the tale of how the deity Ta-aroa hatched from a cosmic egg and made the world from himself and the egg shell.

Even today, some followers of Christianity believe that the heavens and the Earth – the Universe – were forged in seven days, by God. Every culture has nurtured its own religious beliefs, each with its own version of the events that brought the Universe into being. So which one do we believe?

New beliefs

Believers accept the scenarios above on the grounds of faith rather than reason. Anyone can invent such a myth about the Universe and its beginning, without worrying too much whether or not it tallies with the available evidence. To get to the

1. China was one of the first nations to excel at astronomy. Chinese astronomers made the first documented observations of an eclipse of the Sun, in 899 BC.

truth, what we need is a method of thinking that is based on reason and logic. It should let us construct theories about the Universe, which we can then test by comparing the theories' predictions with real observations. In fact, we already have just such a method. It's called science.

The world's first scientist, at least in recorded history, was Thales of Miletus. Thales lived between the 6th and 7th centuries BC in Ionia, on the eastern shores of the Aegean Sea. The Ionians were a practical people, a trait borne out not only by their workers but also by their thinking men. Ionian scholars were in their element poking and prodding the world around them to find out how it worked.

Thales rejected the idea that the heavens and Earth were created by mysterious and unfathomable gods. Instead, he argued that the Universe was shaped by forces of nature, which he believed human beings were capable of understanding. One of his early theories supposed that the Universe is a great sea of water, upon which floats a disc that is the Earth. Using this theory – essentially the first scientific theory of cosmology – he was able to explain earthquakes.

Of course, the theory was wrong. Today, views returned by spacecraft and powerful telescopes show us immediately that the Universe isn't made of water and the Earth isn't a disc. Thales simply lacked the technology to make such observations. Nevertheless, his system of reasoning, comparing theories with experimental observations – the scientific method – was sound. Indeed, our current understanding of the cosmos relies on it.

1

1. Thales of Miletus put forward the first theory of cosmology: the Earth was a flat island afloat in a universe of water and surrounded by air and fire.

2

2. It was Galileo Galilei (1564–1642) whose astronomical observations led to the modern understanding of the Universe.

On the right track

It was some two millennia after the time of Thales before the modern view of the Universe began to emerge. In 1576, English astronomer and mathematician Thomas Digges suggested that the Universe is infinite, and that stars like our Sun are evenly distributed through it. Although not quite correct, it was a big improvement on the prevailing belief of the era – that the stars occupied a great crystalline sphere at the edge of a Solar System that was centred on the Earth.

If Digges made the first crack in the crystalline sphere theory, then it was Italian astronomer Galileo Galilei who finally shattered it. Just half a century after Digges's suggestion, Galileo noted that more stars were visible when he looked through his telescope than when he looked with the unaided eye. Although some of the new stars that he was seeing must be intrinsically fainter, some he reasoned must simply appear fainter because they are farther away. In other words, the stars aren't all at the same distance as required by the crystalline sphere theory.

At around the same time, Galileo caused yet more Establishment upset when he published telescopic observations of the planet Venus. These showed that Venus has phases – just like the phases of the Moon – the pattern of which demonstrated clearly that the planet orbits the Sun and not the Earth. This supported the then controversial (but nevertheless correct) view, put forward by Polish astronomer Nicolaus Copernicus, that the entire Solar System in fact revolves around the Sun.

In 1750, English natural philosopher Thomas Wright built upon Galileo's finding that the ▷▷

MUSIC OF THE SPHERES

The ancient Greek philosopher Aristotle (384–322 BC, right) suggested that the Universe is made from a number of crystalline spheres nested within one another like the layers of an onion. At the centre of the onion sits the Earth. Every other object in the Solar System – the Sun and the other planets – each has its own sphere into which it is set. Each sphere revolves independently, and this carries the celestial objects on their paths across the sky. The outermost sphere was supposed to hold the distant stars, and as such represented the edge of the Universe. Aristotle believed that while everything inside the Universe was made of the four basic elements – earth, air, fire and water – the stars on its boundary were made of a 'fifth' element, known as quintessence.

1

Previous page: The Pleiades, a star cluster about 400 light years from Earth. The cluster is some 50 million years old.

1. In 1543, Nicolaus Copernicus challenged the conventional view with this picture of the planets and the zodiac circling the Sun.

2. An image from *Original Theory of the Universe* (1750) by Thomas Wright. He was the first to suggest that the Milky Way is a disc of stars.

stars were spread across space. He suggested that the Milky Way – the broad, pale band that's visible across the sky on a dark, clear night – is a disc of stars, and that our Sun and Solar System are embedded within the disc. This is essentially the form that our Milky Way galaxy is now known to take. Wright even went so far as to suggest that space could be littered with other similar systems of stars – other galaxies. But it would take until the 1930s for this idea to be confirmed. Before then many astronomers believed that the Milky Way was all there is – a solitary island of stars in a big, empty Universe.

BEYOND THE MILKY WAY

The first pieces of evidence that there were celestial objects lying outside the Milky Way were found by astronomers studying comets – frozen bodies that wander through our Solar System – a field far removed from the study of the Universe at large.

When a comet passes near to the Sun, its surface evaporates into a cloud of steam that reflects sunlight to create a fuzzy glowing patch on the sky. Comet-hunters would search the skies for such fuzzy patches and then observe how the position and brightness of each patch changed

2

Professor Sir Martin Rees, Britain's Astronomer Royal, believes that the cosmological history of the Universe is now more accurately determined than the geological history of the Earth.

as the comet travelled nearer to, or farther away from, the Sun.

But during the 18th century, some astronomers began to find comet-like fuzzy objects that didn't seem to be moving or changing in brightness. These objects became known as nebulas, from the Latin word for 'cloud'. Between 1760 and 1784, French comet-hunter Charles Messier made the first catalogue of these nebulas. There were originally 103 of the so-called Messier objects (this has now grown to 110), designated by their characteristic 'M' numbers, which are still used today.

In the 1860s, astronomers discovered what the nebulas really were. They used the new discipline of spectroscopy – splitting light into its spectrum of colours and measuring the brightness of each colour. The spectrum of light given off by each chemical element has its own characteristic pattern of bright and dark colours. By spotting these

The 18th-century German philosopher Immanuel Kant was an early adherent to the view that distant nebulas might be complete star systems beyond the Milky Way.

1

1. Spiral galaxy M83, one of the original 103 objects identified and classified by French astronomer Charles Messier (1730–1817). M83 lies 20 million light years away and is thought to be very similar to our own Milky Way galaxy.

1. German physicist Josef von Fraunhofer (1787–1826) demonstrates an early spectrometer. Its invention allowed astronomers to tell what distant celestial objects were made from – a vital clue in piecing together the story of the Universe.

patterns in the light from a celestial object, astronomers can work out the object's chemical composition.

When they did this for the nebulas, they found that some had spectra matching that of glowing hydrogen gas, and so were nothing but interstellar gas clouds. Others, however, resembled the more complicated spectra of star light. Roughly two-thirds of the nebulas fell into this latter category, so they had to be groupings of stars. Many of these groupings also displayed intricate spiral patterns, leading to the name 'spiral nebulas'. The question was: did these systems of stars lie inside or outside the Milky Way?

Galaxy quest

Astronomers were divided in their answers. Some had found that there seemed to be fewer nebulas in the plane of our galaxy's disc than in the rest of the sky. This, they argued, suggested a causal link between the nebulas and the Milky Way – evidence that the nebulas were nearby.

On the other hand, some astronomers cited observations of novas – a type of stellar outburst – as evidence that the nebulas must be farther away. All novas have about the same well-known intrinsic brightness. So if you spot the bright flare of a nova in the sky and measure its apparent

The first suggestion that the Milky Way might have a spiral structure, like the spiral nebulas, was made in 1900 by Cornelius Easton.

brightness, you can calculate by how much its light has been dimmed with distance. This then tells you how far away it is. In 1917, four very dim (and therefore very distant) novas were seen in a spiral nebula called Andromeda. They were so faint that astronomers placed Andromeda some 10 million light years from Earth. (A light year is the distance that light travels in a year at a speed of 300,000 km [186,420 miles] per second.) By comparison, in 1918, American astronomer Harlow Shapley deduced that the Milky Way's disc was roughly

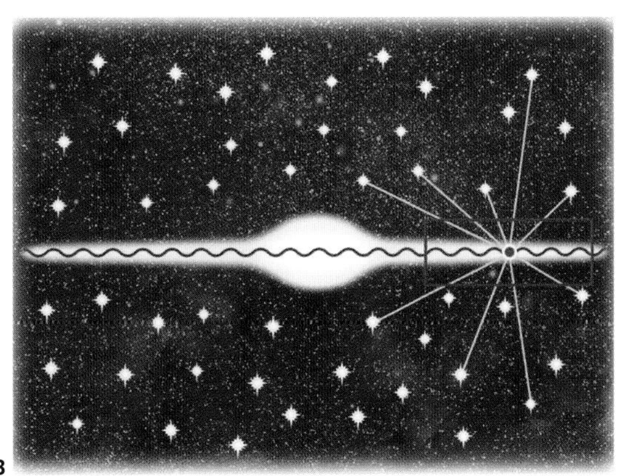

2. The Andromeda Galaxy, M31, the closest galaxy to our own, was crucial in establishing the distance scale across the Universe.

3. Harlow Shapley measured the distances of globular clusters (dots), to deduce that the Sun (at right) is some way from the centre of the Galaxy. But dust in the plane of the Galaxy hides its centre from our view.

100,000 light years across – suggesting that Andromeda lies outside the Milky Way.

The argument was finally resolved by American astronomer Edwin Hubble. Between 1919 and 1924, Hubble used the 2.5-metre (8-ft) Hooker Telescope on Mount Wilson, California, to make detailed observations of the Andromeda nebula and another gathering of stars called the Triangulum nebula. Hubble was searching the nebulas for a particular type of star known as a Cepheid variable.

The first Cepheid variable star – Delta Cephei (after which the species is named) – had been discovered in 1784. Cepheids are special because their brightness varies regularly over a period of days to a few weeks. In 1912, American astronomer Henrietta Leavitt discovered that the period of a Cepheid's variability is linked to the average value of its intrinsic brightness.

1. Edwin Hubble's discovery that the Universe is expanding pointed to the idea that the Universe began in a hot dense state (later called the Big Bang).

2. Hubble measured the distances to nearby galaxies by observing a type of variable star known as a Cepheid. The first Cepheid to be discovered was Delta Cephei, pictured here.

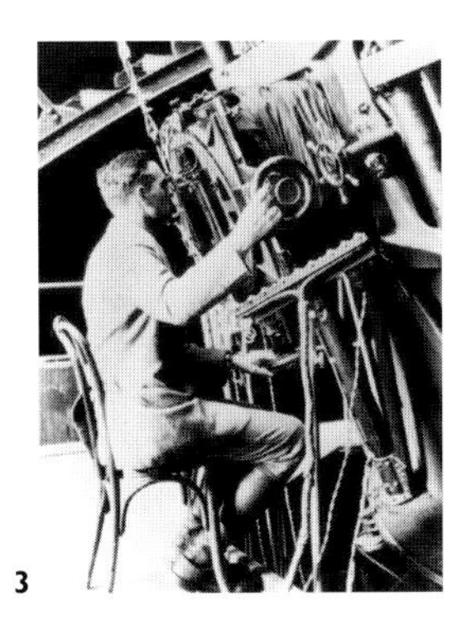

3. Edwin Hubble at Yerkes Observatory. Its 1-m (40-in) refracting telescope is still the largest in the world.

4. Yerkes Observatory where Hubble trained as an astronomer.

EDWIN HUBBLE – BOXER, LAWYER, ASTRONOMER

Best known as the astronomer who discovered that the Universe is expanding, Edwin Hubble (1889–1953)was in fact a man of many talents. His first degree was in law, which he obtained from the University of Chicago in 1911. He pursued his law studies further as a Rhodes Scholar at the University of Oxford, where he also excelled as an amateur boxer and was even offered the chance to turn professional, which he declined. Hubble's interest in astronomy began while studying at Chicago. In 1914, after a brief spell practising law in Kentucky, he started working as an assistant at Yerkes Observatory near Chicago. His work there earned him a PhD in astronomy in 1917. After a period serving in the American infantry during World War I, Hubble joined the staff at Mount Wilson Observatory, California, in 1919, where he would later carry out the groundbreaking work for which he became famous.

1. Henrietta Leavitt found a link between Cepheid variable stars' brightness and their variability.

2. Vesto Slipher noted that many distant galaxies had redshifted spectra and so must be rushing away from us.

This discovery was vital to astronomers trying to understand the cosmic distance scale. Now all they had to do to measure the distance to a spiral nebula was find a Cepheid in the nebula and measure its apparent brightness and its period. As Leavitt had established, the period reveals the star's intrinsic brightness. Comparing this with its apparent brightness shows by how much the light from the star has been dimmed with distance, and therefore how far away it is. This was just what Hubble did. His findings showed indisputably that the Andromeda and Triangulum nebulas lie far outside our Milky Way. He concluded that the spiral nebulas are galaxies of stars, much like the Milky Way, but hundreds of thousands of light years distant.

The expanding Universe

Resolving the distance debate was one of Hubble's early contributions to our understanding of the cosmos. He would soon make one of the most important cosmological discoveries ever.

Hubble's big discovery would extend the work of Vesto Slipher, an astronomer at the Lowell Observatory, Arizona. Between 1912 and 1925, Slipher studied spiral nebulas, looking in particular at the characteristic bright and dark pattern in the spectrum of light from each nebula. He found that of the 45 nebulas he looked at, 43 had spectra that were 'redshifted'. This means that the pattern in the spectrum was still there, but shifted to colours that were redder than normal.

Redshift is caused by a phenomenon known as the Doppler effect, which also explains why, when an ambulance passes you, the pitch of its siren becomes lower. The motion of the ambulance away from you stretches out the siren's sound waves to a lower frequency. The same thing happens in the spectra of the spiral nebulas. The light from their

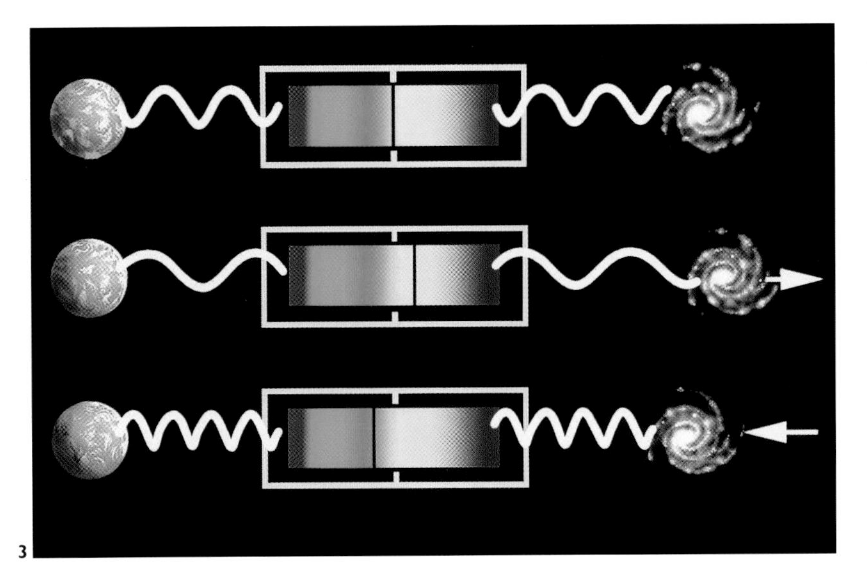

3. The colour of the light from celestial objects is often shifted towards the blue or red end of the spectrum, depending on whether the object is moving towards or away from Earth, respectively.

4. Hubble's Law states that the more distant a galaxy is, the faster it is receding due to cosmic expansion, so the greater its redshift.

stars is being stretched out and shifted towards the low-frequency red end of the spectrum. The reason? Like cosmic ambulances, the nebulas are moving away from us.

Hubble, and his assistant Milton Humason, used the Cepheid method to obtain distances to Slipher's nebulas (now known to be galaxies in their own right). In 1929, they showed that the redshift of each galaxy is simply its distance multiplied by a constant number. The number has become known as Hubble's constant, and the relationship between distance and redshift, Hubble's Law.

Because the Doppler effect relates a galaxy's redshift to how fast it's moving, Hubble's distance-redshift law equates to a distance-speed law: the more distant a galaxy is, the faster it's rushing away. The punchline was that the Universe is expanding.

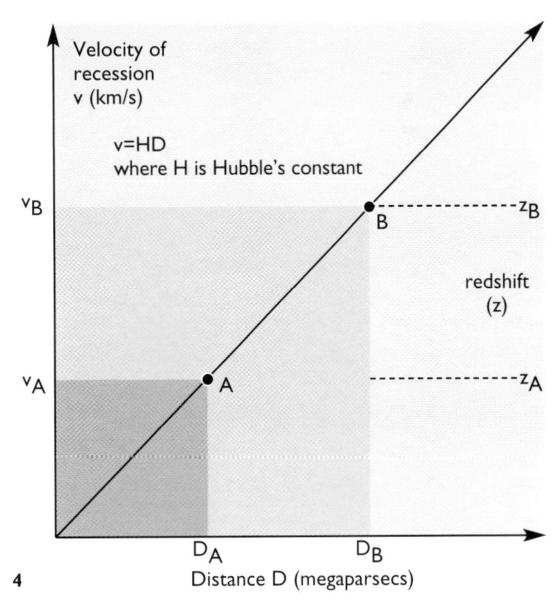

ENTER THE THEORISTS

It was official: space is getting bigger. The ball was now in the court of the astrophysicists to figure out why – to devise a theory that could accommodate an expanding Universe. However, when Hubble and Humason made their discovery in 1929, such a theory already existed, and had done for 14 years. It was Albert Einstein's general theory of relativity.

General relativity was a new theory of gravity, replacing the earlier ideas of Sir Isaac Newton. Einstein believed that space and time – rather than being solid and immutable – are flexible entities. He constructed a complex set of mathematical equations linking their flexibility with the matter that they contained. According to the theory, objects moving in a gravitational field follow curved trajectories simply because space and time are themselves curved. So whereas we normally think of an object moving under gravity – such as a ball thrown up in the air – as following a curved path in flat space, relativity says that the ball is,

1

Empty space is flat and light beams therefore travel in straight lines.

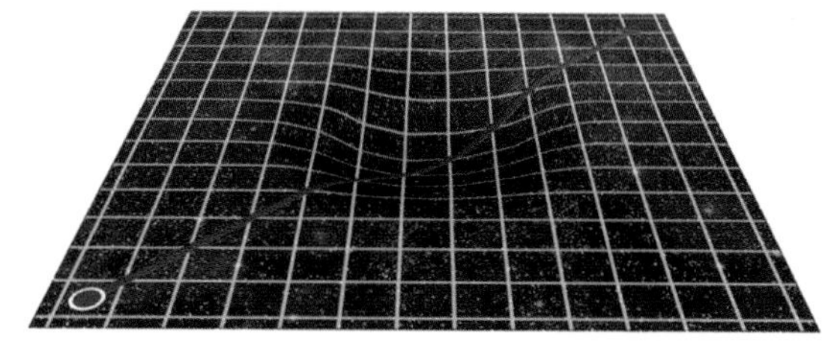

2 The presence of a heavy object, such as a star, distorts space, bending the path of a light beam.

Albert Einstein once said: 'The most incomprehensible thing about the Universe is that it is comprehensible.'

in fact, following as straight a path as it can in curved space. The often-used analogy is to think of space as the surface of a trampoline. Roll a marble across the surface and it moves in a straight line. But place a massive body, such as a cannonball, on the trampoline and it creates a large dent, making the marble follow a curved path instead.

One key consequence of this is that gravity affects not only massive objects, but light beams as well. The theory predicts that a light beam moving close to a massive body should get deflected. And in 1919, during a total solar eclipse, this prediction was verified. With the Sun's glare blotted out by the Moon, astronomers could photograph stars on the Sun's limb (the edge of its disc as seen in the sky), finding their positions to have been shifted ever so slightly as the light rays from the stars were bent by the Sun's gravity. General relativity has since been subjected to other tests and has passed them all, establishing itself as our best gravitational theory.

Boom time

The Universe at large is governed entirely by gravity, so general relativity was a landmark breakthrough for cosmology. Soon after publishing the theory, Einstein himself tried to use it to build a mathematical model describing the Universe. This

1. In 1915, Einstein presented his general theory of relativity – which would provide the mathematical basis for the Big Bang theory.

2. General relativity ascribes gravity to curvature of space and time. Massive objects distort the space and time in their vicinity.

3. In 1919, solar eclipse observations proved that light rays from distant stars are bent as they pass near to the Sun, vindicating general relativity.

3

was in 1917, however, before cosmic expansion had been discovered. The prevailing view then was that the Universe was static. So when general relativity actually predicted that the Universe should either expand or contract, Einstein dismissed the result and instead looked for ways to modify the theory and make it compatible with a static Universe.

But not everyone was so easily deterred. In 1927, a Belgian astronomer and cosmologist, Georges Lemaître, published a solution to Einstein's equations that could describe the behaviour of a range of expanding universes. Before astronomers had officially discovered the expansion of the Universe, Lemaître saw how his solution fitted perfectly with the galaxy redshift observations of Slipher. Others had also arrived at the same result, but Lemaître was the first to realize what it meant.

He grasped that if galaxies were moving farther apart with time, then in the past they must have been closer together. Keep winding back the clock and you come to a time when the galaxies

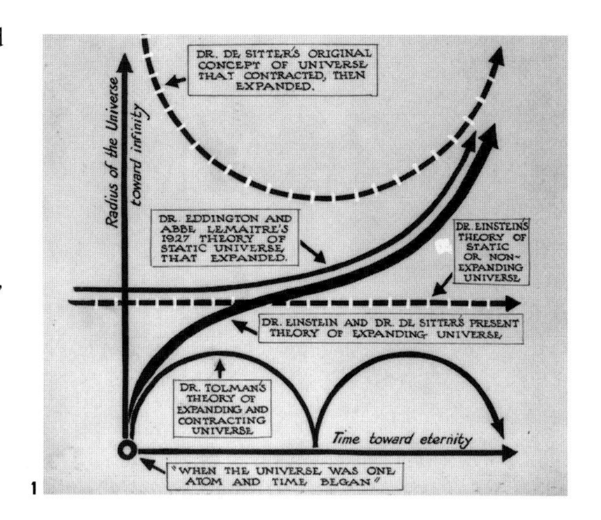

1. Georges Lemaître's 1927 diagrams of his re-collapsing, ever-expanding and oscillating Universe models.

THE ANCIENT UNIVERSE

Edwin Hubble's law for the expansion of the Universe says that the speed at which any galaxy is receding is equal to its distance from our own Milky Way galaxy multiplied by a number known as Hubble's constant, usually denoted by the letter H. Measurements of galactic distances and recession speeds allow the value of H to measured. H tells astronomers how fast the galaxies are rushing apart, enabling them to calculate how much time has elapsed since the Big Bang, when the galaxies were all on top of each other. This is the age of the Universe. Current observations put H at around 60 km (37 miles) per second per million parsecs (a parsec is a unit of astronomical distance equal to 3.26 light years). This gives an age for the Universe of about 15 billion years.

2. Sir Arthur Eddington arranged for Lemaître's work to be published in *Monthly Notices of the Royal Astronomical Society.*

3. Sir Fred Hoyle, a great opponent of the Big Bang theory. Ironically, he gave it its name.

overlapped. Go back further still and you reach a point where all the matter in the Universe was packed into a small, hot, dense sphere, which Lemaître referred to as the 'primeval atom'. He supposed that the Universe began when this atom, which he envisaged to be about 30 times the size of the Sun, spontaneously exploded. Matter and radiation spewed outwards from the explosion, as did the very space and time that underpin them.

Spread the word

When British astrophysicist Sir Arthur Eddington learned of Lemaître's work – at about the same time that Hubble and Humason were announcing their observations of cosmic expansion – he immediately realized its importance. He arranged for Lemaître's original research paper, which had appeared in an obscure Belgian journal, to be published in the British journal, *Monthly Notices of the Royal Astronomical Society*, where it appeared in 1931.

Lemaître's Universe was the first recognizable incarnation of cosmology as it is presently understood. The theory was developed further throughout the 1930s and '40s, extending its remit back to a time when the initial cosmic seed was increasingly smaller, hotter and denser.

A title for this bold new picture of the cosmos was supplied in the 1940s, ironically by one of the theory's leading opponents: Sir Fred Hoyle, a British astronomer then working at the University of Cambridge. Hoyle refused to believe that the Universe had been born in what he mockingly termed a 'Big Bang'. And the name has stuck.

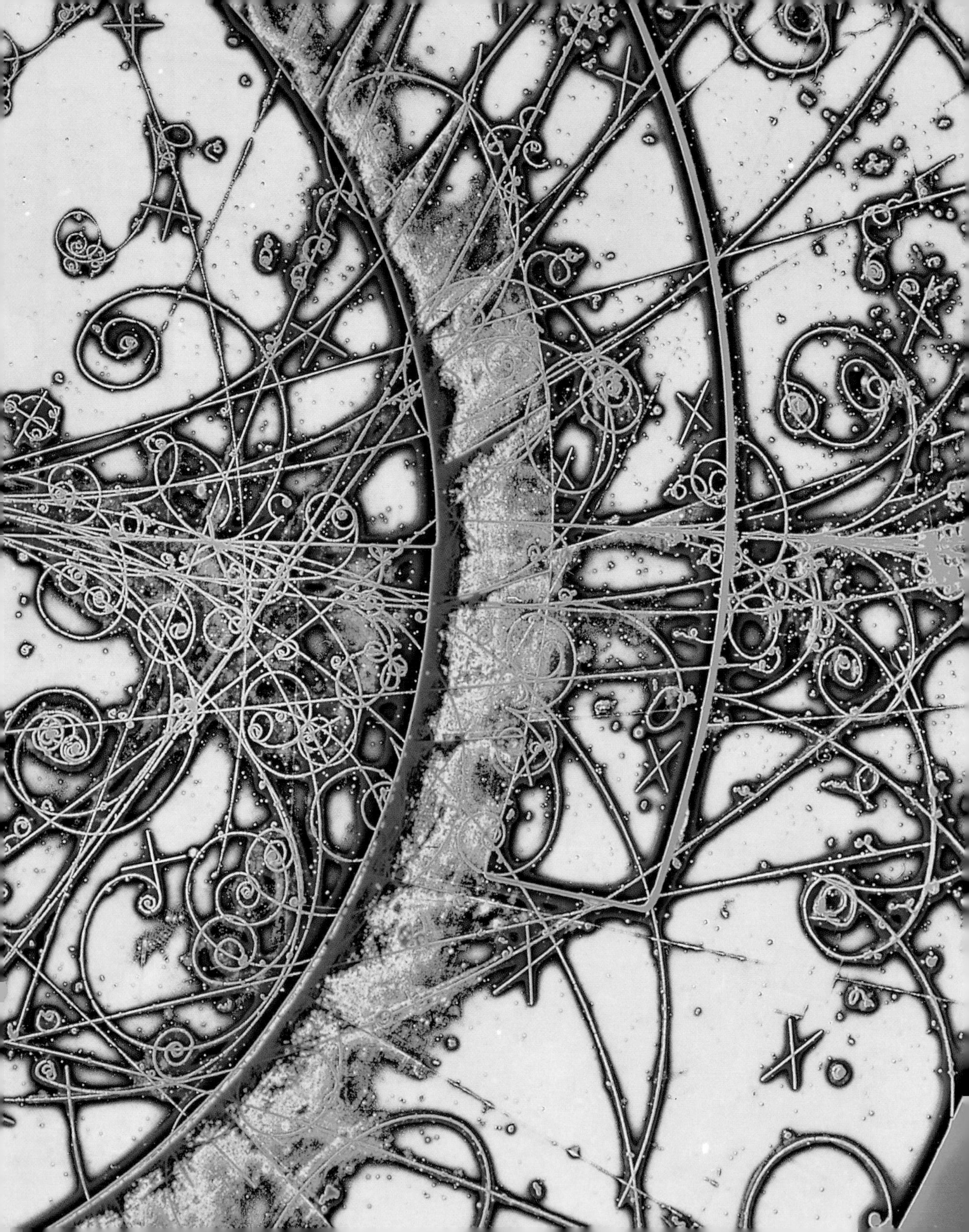

2

THE MODERN UNIVERSE

2

THE MODERN UNIVERSE

With the foundations of the Big Bang theory in place, cosmologists could begin filling in the fine details. They had realized that winding the expansion of space backwards in time meant that today's very large Universe must have started out very small. And so they turned to the science of the very small – quantum physics – to explain how it worked. This told them a great deal about the very early Universe. But one thing in particular was still niggling. How could the magnificence of the cosmos – all those billions upon billions of stars and galaxies – have been born out of nothing, as the Big Bang theory seemed to suggest? Quantum physics was to hold the answer to that question as well. And it would lead scientists to some bizarre possibilities for our Universe – that it may have been created from nothing, and that during its earliest moments parts of it expanded away from each other faster than the speed of light.

Previous page: Subatomic particle tracks recorded at the European particle accelerator lab, CERN. As scientists came to understand more of the early Universe, they realized the relevance of quantum theory and particle physics in its behaviour.

CREATION FROM NOTHING

Roughly 15 billion years ago, the Universe popped into existence where a microsecond earlier there had been absolutely nothing. All the matter and energy that make up everything we see, both in the world around us and – thanks to powerful telescopes – on the other side of the Universe, were created in that single brief instant. But how?

The question has perplexed philosophers over the centuries. How can something emerge from nothing? It seems to violate common sense. In fact, it seems to violate a scientific principle known as the conservation of energy, which says that mass and energy cannot be created or destroyed. With no obvious way to explain creation in terms of what is natural, many turned to the supernatural, claiming that creation must be the handiwork of a creator.

1. The Hubble Deep Field South, a deep view of the sky returned by the orbiting Hubble Space Telescope. The deep field, one of our most distant optical views of space, shows the Universe as it was very close to the time of the Big Bang.

Divine intervention?

History is replete with examples of gods being invoked to explain the unknown. But in most cases science eventually provides a more plausible explanation. For instance, the inhabitants of ancient Scandinavia were unable to explain the phenomenon of thunder, so they concluded that it must be the work of a thunder god – whom they called Thor. We now know that thunder is caused by lightning. Electrical charges build up in clouds and, when these are big enough to overcome the electrical resistance of the intervening air, the charge flows to the ground as a lightning arc. The temperature of a lightning arc can reach 30,000° C, which flash-heats the surrounding air, causing it to expand in a compression wave that we hear as a thunderclap. There's no need for Thor.

Creationists believe that a divine creator brought the Universe into being. Often, they cite a theological principle called the first cause argument to support their belief. This says that everything in the Universe must have had a prior cause. Either there is an infinite regress of causes, extending for ever into the past, or there was a first cause – a creator.

But this argument is shortsighted. If there exists a creator, then what created the creator? And so on. Deciding, *ad hoc*, that the creator requires no creator is groundless. Why not simply say that the Universe requires no creator? The first cause argument is no more an argument for God than thunder is an argument for a thunder god.

A clockwork Universe

Other creationists have taken on board a scientific principle to help them argue for the existence of God. It's called the second law of thermodynamics. The law essentially says that although energy is conserved, its ability to make things happen – or, in scientific language, 'do work' – is steadily dwindling away. For the Universe, it means that the stars and

MOMENT OF CLARITY

Albert Einstein realized that the Universe could be created from nothing during a conversation with physicist George Gamow. While the two men were out walking in Princeton, New Jersey, one day during the 1940s, Gamow mentioned how one of his students had calculated that it's possible to make a star from nothing because its mass energy is exactly equal but opposite to its gravitational energy. Einstein, realizing immediately that the same principle could apply to the Universe at large, stopped in his tracks. He and Gamow were crossing a road at the time, forcing several cars to stop in order to avoid hitting them.

1. A computer simulation, modelling the processes by which galaxies and clusters of galaxies formed during the early Universe. The region shown is 30 million light years across.

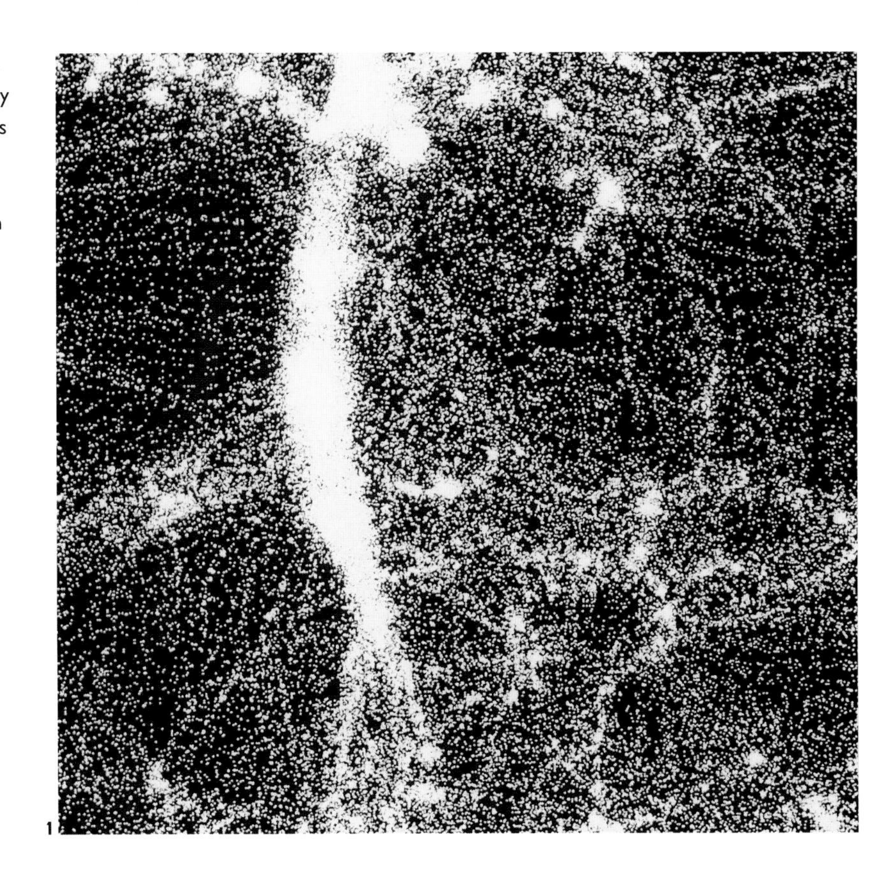

galaxies will eventually burn out and die, leaving nothing but a black, gloomy void in their place. This scenario is called the 'heat death' of the Universe. It's as if the Universe is like a clockwork toy, steadily unwinding and destined to grind to an eventual halt. If the Universe is unwinding, argue the creationists, then someone or something must surely have done the winding up.

This argument is also groundless. The Universe started life in a 'wound-up' state simply because it was born small and dense, and therefore hot. Matter and radiation have a greater capacity to do work when they are hot than they do when they are cold. For example, in a car engine, each piston rod moves from one end of its respective cylinder, where the gas (an ignited petrol-air mixture) is very hot, towards the other end where the gas (cold air) is much cooler. When the ignition is switched off, both ends of the cylinders are then cold, and the engine stops. And that's why the hot young

Universe was wound up. At a temperature of a thousand billion billion billion°C, it had a prodigious capacity to do work.

So if God didn't make the Universe, what did?

The quantum cosmos

The best explanation we have for the origin of the Universe comes from the science of subatomic particles – quantum theory. This branch of physics was developed in the early 20th century to explain the behaviour of atoms and radiation.

Quantum theory makes some strange predictions. One of the most bizarre was discovered in 1927, by German physicist Werner Heisenberg. It's called the uncertainty principle, and it says that you can never know both the speed and the position of a subatomic particle at the same time: measuring one quantity diminishes the precision with which you can know the other. This is not due to any inaccuracy of experimental measuring devices; it is inherently forbidden by quantum physics. No one really understands why Heisenberg's principle operates in this way. But the principle is an 'empirical' law: its predictions agree well with the results of experiments, which is what matters; and it is nowadays taken as cornerstone law that underpins much of quantum physics.

Heisenberg went on to show that linking speed and position like this is equivalent to linking mass

1. German physicist Werner Heisenberg discovered the so-called uncertainty principle, a cornerstone of quantum physics and key to explaining how the Universe was created from nothing.

Opposite: Computer simulation of subatomic particle tracks, as are regularly produced in the giant atom-smashing particle accelerator at CERN in Switzerland.

and time in the same way – in the quantum world particles can pop in and out of existence so long as their masses and the times that they live for before disappearing again obey the uncertainty principle. Empty space is then no longer empty, but is in fact a seething mass of particles flitting in and out of reality. The effect is real. Evidence for these so-called virtual particles has been seen in the attractive force that they exert on two metal plates placed close together (the Casimir effect) and in the shift they produce in the frequency of radio waves given off by hydrogen atoms (the Lamb shift).

During the early 1970s, physicist Edward Tryon, of the City University of New York, suggested that just as quantum uncertainty brings virtual particles into existence, it could also have given birth to the tiny seed from which our Universe grew.

Although matter is being created from nothing in this scenario, the principle of conservation of energy is obeyed. Astronomical observations show that our Universe is balanced so that the energy associated with its mass is equal but opposite to the energy locked away in its gravitational field. Thus, the net energy of the Universe is zero – no energy has been created or destroyed.

However, there's just one problem. Packing the Universe into something the size of a quantum particle creates a gravitational field so strong that it should have crushed the Universe out of existence in an instant. For the idea to work, something must have happened to the embryonic Universe to expand it out of the quantum realm before it re-collapsed. But what? The answer to this question eventually emerged in the early 1980s.

IS THE UNIVERSE ITS OWN MOTHER?

The most likely route by which the Universe came into being is quantum creation from nothing: the Universe popped into existence just as subatomic particles pop into existence in accordance with quantum uncertainty and Heisenberg's principle. Yet some scientists still hold that the Universe was created from something. One version of this idea supposes that the 'something' was the Universe itself, and that creation took place via a loop in time. Richard Gott III and Li-Xin Li, both at Princeton University, New Jersey, were the first to suggest the idea, in 1998. It relies on an earlier theory that our Universe keeps giving birth to baby universes (right), which branch away and expand to form new universes in their own right. Gott and Li wondered what might happen if one of these baby universe branches could loop back on itself through time, rejoining the main trunk of our Universe at the point we call the Big Bang.

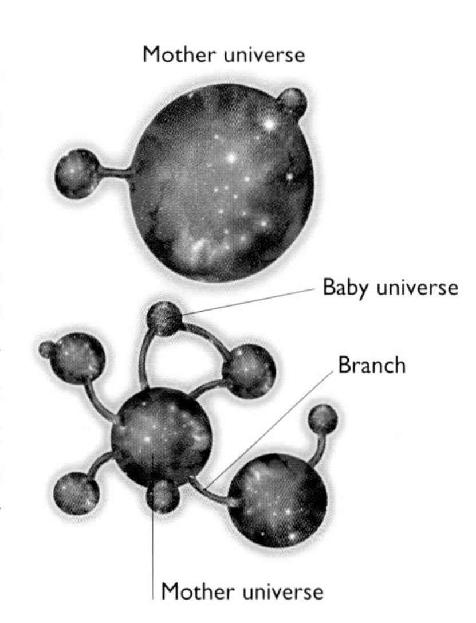

INFLATION

The theory of cosmological inflation says that just one hundred-million-billion-billion-billionth of a second after it was born, the Universe underwent a massive growth spurt. This expansion blasted the Universe up from the quantum world, before slowing down into the more sedate expansion that we see today. Astrophysicists introduced inflation to iron out a number of minor bugs with the standard Big Bang theory. As well as deepening our understanding of how the Universe at large came to be, inflation also explains how smaller-scale structures – the galaxies – were created.

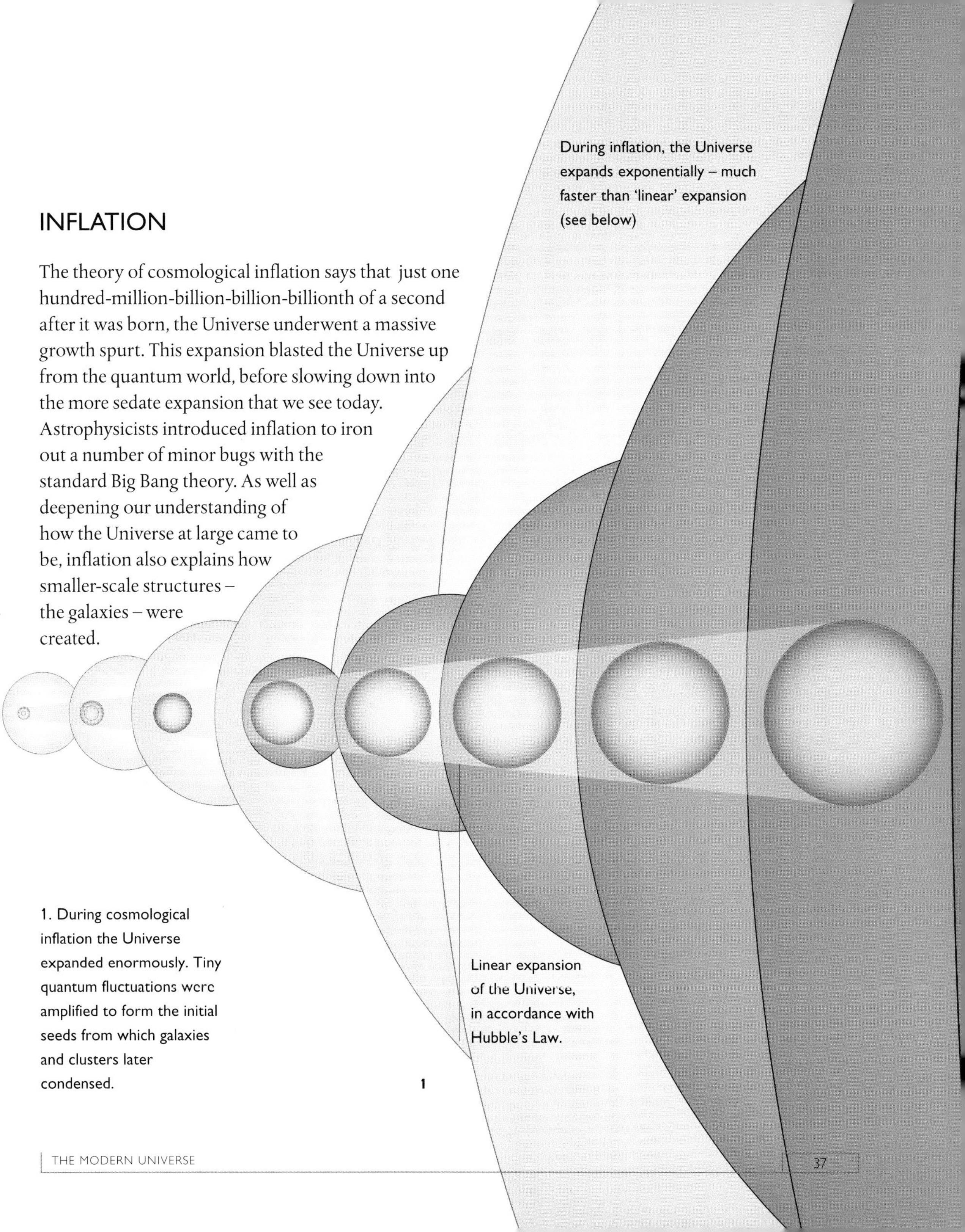

During inflation, the Universe expands exponentially – much faster than 'linear' expansion (see below)

Linear expansion of the Universe, in accordance with Hubble's Law.

1. During cosmological inflation the Universe expanded enormously. Tiny quantum fluctuations were amplified to form the initial seeds from which galaxies and clusters later condensed.

1

Early days

The first work on what came to be known as inflation was carried out by a Russian physicist, Erast Gliner, of the Institute for Physics and Technology in Leningrad, in 1965. Gliner envisaged a rapid phase of expansion in the very early Universe – but the idea never caught on. Nearly 15 years later another Russian, Alexei Starobinsky, then working at the Landau Institute, Moscow, arrived at much the same idea. Starobinsky had been investigating how to marry quantum physics with the laws of gravity. The result, quantum gravity, applies to tiny objects with very strong gravitational fields – just like the early Universe. Physicists are still, today, searching for the correct theory of quantum gravity, but Starobinsky's calculations, in 1979, showed that super-rapid expansion of space was one of its consequences.

Alas, the communication restrictions imposed by the Soviet Union meant that Starobinsky's work went unnoticed by Western scientists. But just two years later, Alan Guth, of the Massachusetts Institute of Technology, published a scientific paper outlining a theory in which particle physics processes could make the early Universe expand at a colossal rate. Guth called the theory 'inflation'.

1

During inflation – the rapid expansion of the very early Universe – the Universe grew by a factor of 10 to the power 70 – that's a 1 followed by 70 zeros.

The mechanism for inflation that Guth envisaged relies on a type of subatomic particle, called the Higgs boson, that is believed to pervade all of space. Just as steam undergoes what are called phase transitions, changing it into liquid water when its temperature gets low enough, the Higgs particles underwent phase transitions as the very early Universe expanded and cooled below certain threshold temperatures. During a phase transition, all the Higgs particles in the Universe dropped from a high-energy state to one of lower energy. But this would not have occurred at exactly the same time in all regions of the Universe. In some places it would have happened slightly earlier, and in others slightly later. So in some places, small pockets of Higgs particles would have remained in a high-energy phase, while all the particles around had slipped down to low energy. In this case the pocket, and the very space that contained it, would have inflated rapidly – rather like an expanding bubble of steam in a pan of boiling water. Eventually, the pocket would fall into the lower energy phase, at which point inflation ended. This was the essence of Guth's theory.

1. Recent surveys of the large-scale Universe indicate that galaxies and clusters lie on the surfaces of giant bubbles in space. The empty bubbles are known as 'voids'.

2. A computer simulation of the elusive Higgs boson, showing its decay producing four subatomic muon particles (yellow tracks).

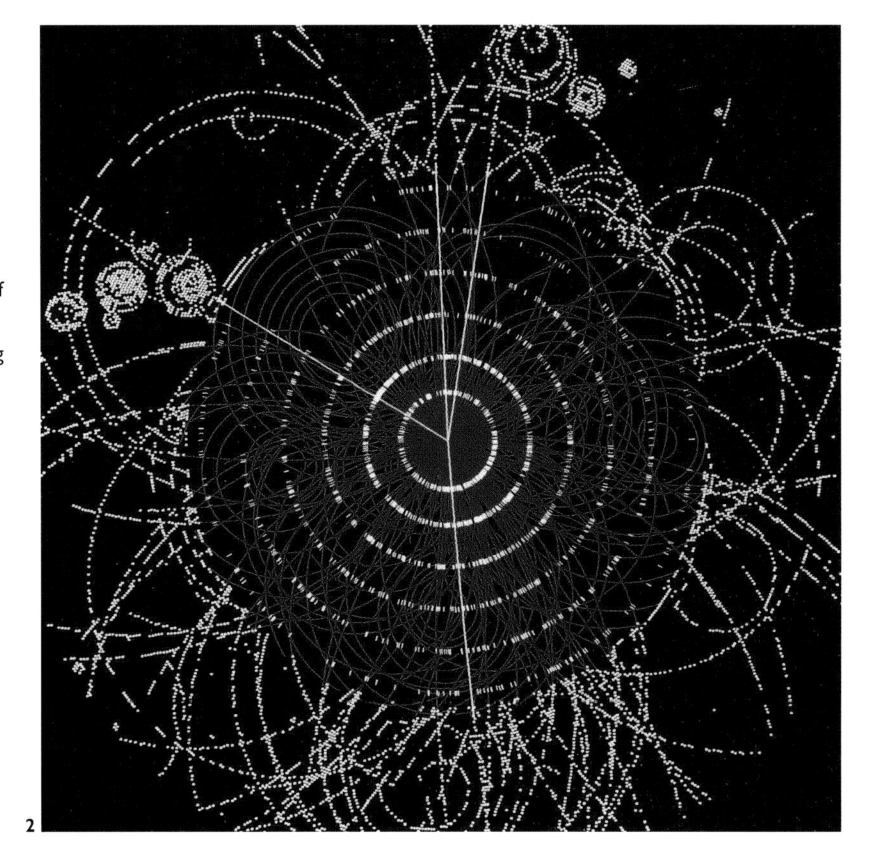

2

Tiny acorns and mighty galaxies

Some critics of inflation initially argued that the rapid cosmic expansion would make it impossible for galaxies to form in the Universe. Galaxies were formed by gravity. Tiny irregularities in the density of matter throughout space grew gravitationally, attracting more matter to become big irregularities. The process snowballed until the irregularities were very big indeed – galaxy-sized. The critics worried that inflation would smooth out the irregularities so much that there was no way they could have grown into galaxies within the age of the Universe. You can imagine space prior to inflation as rather like a wrinkled blanket. Inflation is like stretching the blanket out tight, making all the wrinkles – the irregularities – disappear. Or so the critics argued.

In fact, the situation could hardly have been further from the truth. Inflation may smooth the soil, but it also plants its own seeds from which the galaxies will eventually grow. It's another startling consequence of Heisenberg's uncertainty principle: quantum theory not only explains the birth of the Universe but the birth of the galaxies as well.

It works like this. Quantum fluctuations put the sea of Higgs particles, which are causing the Universe to inflate, in a state of constant flux. Virtual Higgs particles are popping into existence and then quickly disappearing again in accordance with the uncertainty principle. Virtual particles are created in pairs – a particle and an antiparticle – which recombine when it's time for the pair to disappear. But when space is inflating, pairs of virtual particles are dragged apart before they have time to recombine. Rather than disappearing,

THE HIGGS BOSON

Particle physicists believe that all the matter in the Universe is given its mass by the action of a single type of subatomic particle that was was created from the energy that drove inflation. The particle is known as the Higgs boson, after Professor Peter Higgs (right) of the University of Edinburgh, who postulated its existence in 1964. Space is thought to be pervaded by a sea of Higgs bosons. The sea produces a drag effect on other particles moving through it, which shows up as inertia – the resistance to motion that defines mass. Recently, scientists at the CERN particle physics laboratory, near Geneva, caught what they believe to be the first sniff of the Higgs particle at work. An experiment with CERN's Large Electron Positron Collider produced a shower of subatomic particles of types matching those given off when a Higgs boson decays.

1. Ripples in the cosmic microwave background radiation, as seen by astronomers at the Mullard Radio Astronomy Laboratory, Cambridge. The ripples were imprinted by the same irregularities in the density of matter in the Universe from which the galaxies later formed.

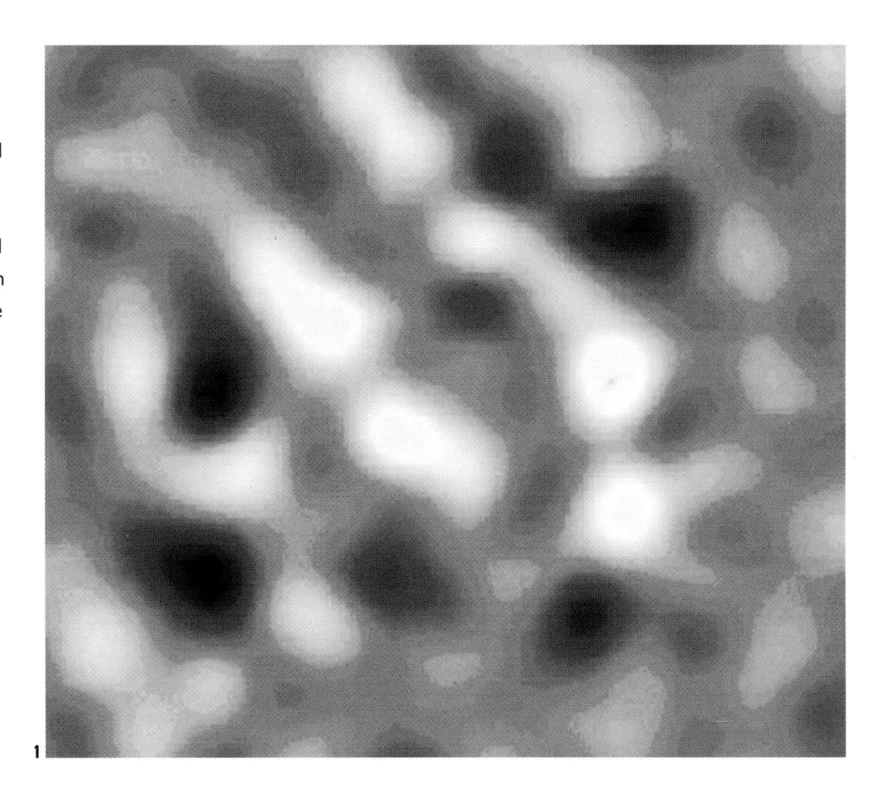
1

quantum fluctuations are then said to 'freeze in'. And it's these frozen-in fluctuations – blown up to cosmic-length scales by inflation – that formed the initial irregularities from which the galaxies grew.

The final piece

These pre-galactic irregularities were imprinted upon the cosmic microwave background radiation – the diffuse electromagnetic echo left behind by the Big Bang and which still pervades the Universe today. Observations of the microwave background, by spacecraft and high-altitude balloons, have shown that it contains irregularities of exactly the form predicted by inflation.

Many variations on the standard inflationary paradigm now exist. Choosing between them will require more detailed observations of the background radiation, which are expected within the next few years. Nevertheless, evidence for the basic idea of inflation is near compelling. Armed with that, we are now in a position to sketch a fairly accurate picture of the history of the Universe – the Big Bang theory as it is presently understood.

THE BOOK OF THE COSMOS

The history of the Universe is a 15-billion-year tale of how one cosmic thing led to another. It is a credit to the human intellect that we have even figured out the length of the story, let alone the title of each chapter and the details of what happened in all but the earliest ones. The story goes something like this…

Time = 0: The Big Bang

Quantum uncertainty brings the Universe into existence as a superdense muddle of space, time and energy.

Time = 10^{-43} seconds: The Planck era

Space and time exist as hazy entities, ruled by the laws of quantum physics. The four fundamental forces of nature – gravity, the strong and weak nuclear forces and electromagnetism – exist as a single unified superforce. At the end of the Planck era (named, incidentally, after quantum physicist Max Planck), the first cosmic phase transition takes place: gravity breaks away from the superforce to become a distinct entity, and space and time become well defined. Physicists are still searching for the correct theory of quantum gravity that governs the Planck era, so very little else is known about the Universe at this time.

Time = 10^{-35} seconds: Inflation

The Universe undergoes the second major cosmic phase transition. The strong nuclear force now splits away from the unified superforce. Only electromagnetism and the weak nuclear force remain tied together, as the so-called electroweak

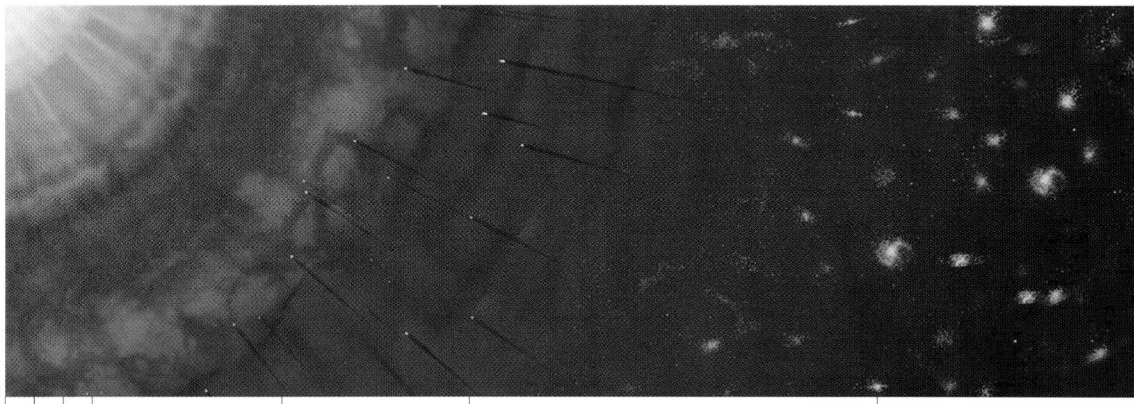

1 The Big Bang (Time = 0)
Inflation (Time = 10^{-35} seconds)
Creation of matter and radiation (Time = 10^{-32} seconds)
Formation of pre-atomic particles (Time = 0.0001 seconds)
Formation of atomic nuclei (Time = 100 seconds)
Formation of atoms (Time = 500,000 years)
Galaxy formation (Time = 1 billion years)

force. This phase transition is thought to have given rise to inflation – the super-rapid expansion of space, which prevented the embryonic Universe from re-collapsing and which also seeded the eventual formation of galaxies.

Time = 10^{-32} seconds: Radiation and matter created

The inflationary phase transition ends, and the energy that drove inflation is transferred to the multitude of Higgs particles that litter the Universe. Many of these particles now decay, releasing the energy as radiation. Since the rapid inflationary expansion had cooled the Universe substantially, this energy release is sometimes referred to as 'reheating'. Soon after reheating, quantum processes cause the radiation itself to decay spontaneously into subatomic particles of matter and antimatter. Most of these particles and antiparticles simply recombine and annihilate, turning themselves back into radiation. However, a small imbalance in the laws of physics produces slightly more matter than antimatter. For every billion particles of antimatter, there are a billion and one particles of matter. Therefore, at the end of this process, a small excess of matter remains, which goes on to make up the material content of the Universe.

Time = 10^{-10} seconds: Electroweak phase transition

The electromagnetic and weak nuclear forces finally go their separate ways. All four forces of nature – gravity, electromagnetism and the strong and weak nuclear forces – now exist as distinct entities.

Today (Time =15 billion years)

1. The story of the Big Bang can be broken down into a number of definite epochs. As the Universe expanded and cooled, the high-temperature laws of physics that governed its earliest moments gave way to their low-temperature counterparts, in a succession of events that physicists call 'phase transitions'.

Time = 0.0001 seconds: Formation of pre-atomic particles

Particles called quarks, which formed shortly after inflation, now bunch together to form more familiar protons and neutrons – the building blocks of atomic nuclei.

Time = 100 seconds: Synthesis of light atomic nuclei

The temperature of the Big Bang fireball drops enough to allow protons and neutrons to stick together without getting ripped apart again by radiation. The nuclei of the lightest chemical elements – hydrogen, helium and a small amount of lithium – are formed.

Time = 500,000 years: Formation of atoms

The temperature is now low enough for atomic nuclei forged during the first 100 seconds to capture electrons and so form the first complete atoms. Up to this time, electrons were constantly scattering radiation this way and that. But as the electrons all suddenly become bound up in atoms, the process ceases. That's why this point in cosmic history is sometimes known as 'last scattering'. With nothing to stand in its way, radiation then streams freely through space. Radio astronomers see this radiation today as the cosmic microwave background – a perfectly preserved fossil relic of the Universe aged 500,000 years.

☆ The modern incarnation of the Big Bang theory reliably describes the Universe from 0.0001 of a second after the moment of creation.

Time = 1 billion years: Galaxy formation

Between 500,000 years and about 1 billion years after the Big Bang, the Universe is in what has been

1

termed the cosmic dark ages. The energy of radiation has dropped to below what is visible with the human eye. And matter is yet to condense into stars and galaxies, making the Universe appear to a hypothetical observer as a bleak, dark and empty place. After about a billion years, however, gravity brings the first generation of stars and galaxies into being. The first lights in the cosmos in 9.5 million years begin to switch on. It must have looked rather like an aerial view of a city waking up before dawn. At first a few faint glimmers appeared. Other lights gradually followed, before the trickle became a flood, and the Universe suddenly lit up in a blazing glare of starlight.

Time = 15 billion years: Today

The Universe is still in the so-called stelliferous era – the age of the stars, also known as the bright ages. And it's set to remain there for the next 10,000 billion years.

1. At about a billion years after the Big Bang, the Universe had cooled down enough for galaxies to start to form. Galaxies come in many shapes and sizes. Shown here is a spiral galaxy known as NGC 4603, which lies 108 million light years away in the constellation Centaurus.

COSMIC CONUNDRUMS

Many people find that some aspects of the Big Bang theory are at best counter-intuitive, at worst utterly baffling. They are in good company. Scientists themselves have yet to comprehend the history of the large-scale (that is millions to billions of light years) Universe in its entirety. However, there are simple answers to some of the most frequently asked cosmic questions.

Where does space end?

It is easy to wonder, when you're looking up at the grandeur of the night sky: Just where does the Universe end? And if it does have an end, or more accurately an 'edge', then what lies outside or beyond that?

In fact, both these questions contain a misconception rooted in the notion that there exists an 'outside' to the Universe. There isn't – at least not one that we know of. Our Universe is defined as the totality of the three dimensions of space (and one of time) that we live in. There is no outside to the Universe because there is no space there. Without space the whole concept of locations, such as 'inside' or 'outside', has no meaning.

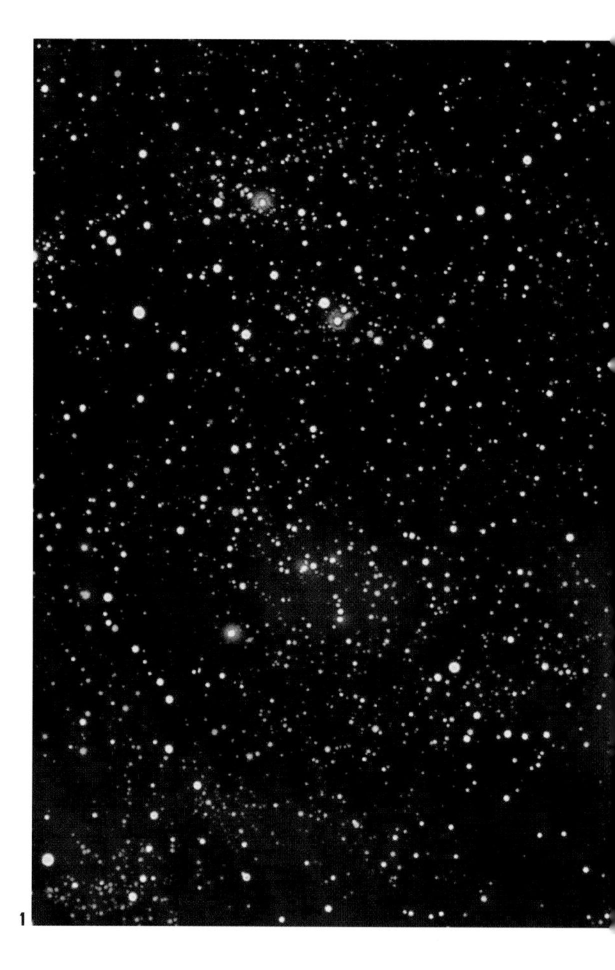
1

You can get a feel for why by imagining a reduction in the number of dimensions that we live in. Say, for example, our Universe had only one space dimension. Then – while you could talk about locations forwards or backwards from your present position – up/down and left/right would cease to have any meaning. Lose that one dimension and forwards/backwards then goes as well.

This is closely related to another often-asked question: if the Universe is expanding, then what is it expanding into? The question stems from our everyday experience in which an expanding object, such as a balloon being blown up, has to expand

1. The Universe is a spectacular place indeed, as the starfield in this image makes clear. But does it continue forever? If a hypothetical space traveller kept going for long enough would that person reach the Universe's edge? Some experts believe that the Universe may be curved back on itself, like a giant sphere. However, the sphere of the Universe would be three-dimensional, so that whereas we exist 'on' the 2-D sphere of the Earth, we are 'within' the 3-D sphere of space.

into something else, such as the surrounding air. But the Universe isn't just a material object expanding within space. The Universe includes all of space – there's nothing outside it to expand into. Space is just getting bigger, and that's it.

But if that's the case, where do you go when you reach the edge? It's possible that space doesn't have an edge at all. On the largest scales the Universe could be a sphere, rather like the Earth. Of course, the Universe wouldn't be a 2-D sphere like the surface of the Earth. Our 3-D space would instead wrap around to form what's known as a 3-sphere. Whichever direction in space you then travelled in, you would eventually come full circle back on yourself.

Why don't we expand?

If space is getting bigger, why don't we see galaxies, planets and even people getting stretched out by the cosmic expansion? Physical objects in the Universe are usually held together by internal forces of some kind. For example, galaxies and stars are held together by the force of gravity. People are held together by interatomic and intermolecular chemical forces. Planets are held together by the same chemical forces plus gravity. And it's these forces that prevent physical objects from getting stretched out by the expansion of space.

It's interesting to note that cosmic expansion is such a large-scale effect that even if small-scale objects did expand, they'd be stretched only by minuscule amounts. For example, Hubble's Law implies that your feet would recede from your head by just 4 billionths of a billionth of a metre per second. That means that in your entire lifetime, expansion of the Universe would make you only about 10 millionths of a millimetre taller.

1

1. Artist's impression of the expanding Universe. But if space is expanding, why don't we and all that we see also get stretched out?

BEFORE THE BIG BANG?

Some people wonder what happened before the Big Bang. What was there before our Universe came into existence? In fact, such a question is ill posed – it makes little sense to talk about what went on 'before' the moment of creation. The reason why is because the Big Bang was not only the beginning of the three dimensions of space; it was also the beginning of time itself. It's impossible to imagine a time before the Big Bang because time simply would not have existed then. Some cosmologists have likened the situation to asking what lies north of the North Pole – a point on the Earth's surface where the only direction you can travel in is south.

Are we at the centre of the Universe?

If distant galaxies are all rushing away from us, regardless of which direction we look in, then surely that means we are at the centre of the Universe? No, it doesn't. Every chunk of space is expanding at the same rate, so the volume of space between us and galaxies lying in every possible direction on the sky is increasing. The classic analogy used to visualize the effect is to draw dots on the surface of a deflated balloon. Now pretend these are the galaxies and slowly blow up the balloon. Pick on any one dot, and you'll see that all the other dots move away from it as the balloon is inflated. However, no one dot is in any way central.

How loud was the Big Bang? Not very loud at all. It was, in fact, completely silent. The misconception that the Big Bang was loud has arisen through the analogy, often made, of likening it to a bomb going off.

When a bomb explodes, a compression wave expands outwards from the explosion, travelling through space to arrive at your ears as a loud bang. But in the Big Bang that created the Universe, there was no compression wave speeding through space. Instead, space itself, along with matter and radiation, was created in the Big Bang.

The matter and radiation then expanded simply because they were being swept along by the subsequent expansion of space. The whole process was very smooth and very quiet.

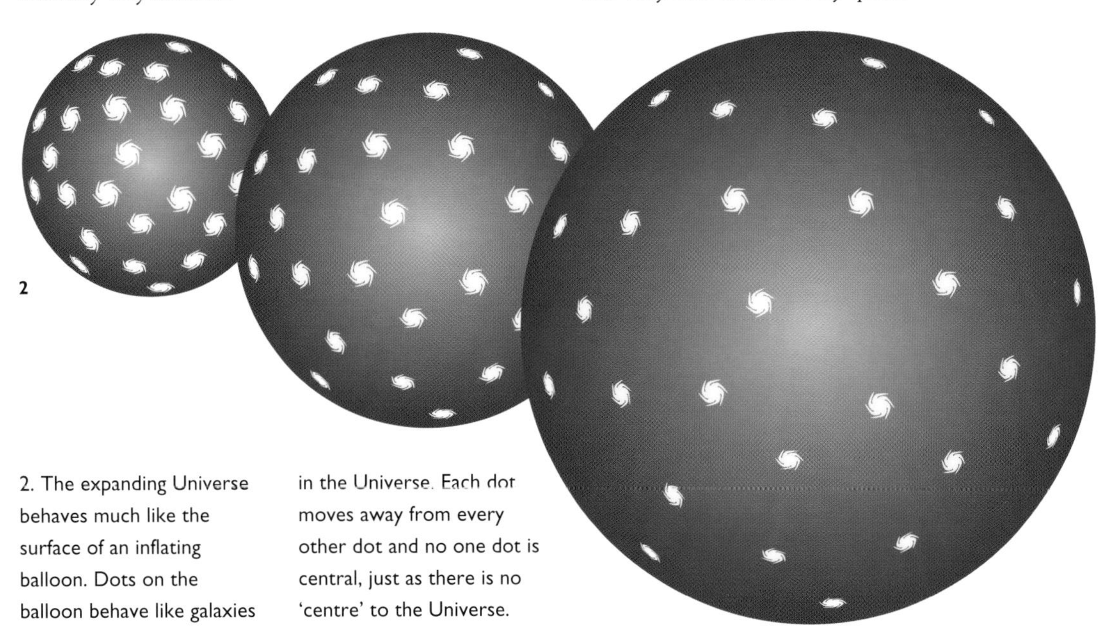

2. The expanding Universe behaves much like the surface of an inflating balloon. Dots on the balloon behave like galaxies in the Universe. Each dot moves away from every other dot and no one dot is central, just as there is no 'centre' to the Universe.

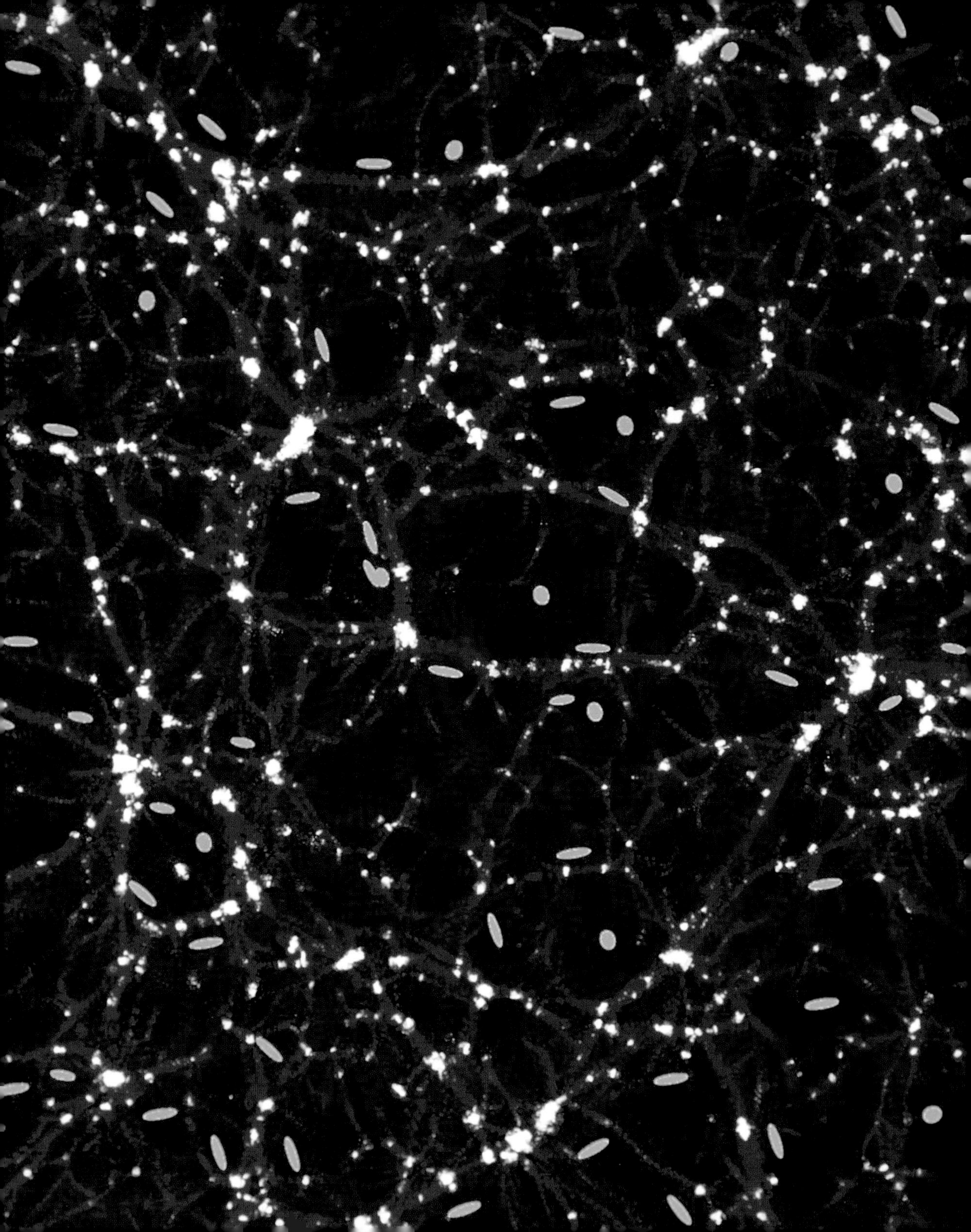

3

TRIUMPHS AND SETBACKS

3 TRIUMPHS AND SETBACKS

The Big Bang theory has far-reaching and, in many cases, mind-boggling implications. It says that the Universe and the galaxies were born from exotic quantum froth, and that space and time are constantly stretching out like some giant four-dimensional piece of elastic. But how do we know that the theory is correct? The cosmos is not something that scientists can experiment on in a laboratory. Instead, it's up to astronomers to gather the clues about how the Universe really works by vigilantly monitoring the night sky. The Big Bang theory is upheld by two cornerstone astronomical observations: the abundances of the light chemical elements, and the temperature of the cosmic microwave background radiation – the electromagnetic echo of the Big Bang fireball. Yet the theory still has its problems. First and foremost, astronomers are still at a loss to explain what 95 per cent of the Universe is really made of.

Previous page: A map of the unseen 'dark matter' in the Universe, produced at the Paris Astrophysics Institute after studying the gravitational motions of galaxies and galaxy clusters.

THE MICROWAVE BACKGROUND RADIATION

When the Universe was roughly 500,000 years old, and its temperature had dropped to some 6000°C – about the same as the surface of the Sun – matter and radiation separated out from one another. Electrons – negatively charged subatomic particles – joined with atomic nuclei that had formed during the first 100 seconds of cosmic history to create the first whole atoms. Before this time, the temperature had been so high that any atoms forming were ripped apart almost immediately by the intense radiation. As electrons are electrically charged, they scatter electromagnetic radiation very well. Atoms, however, are electrically neutral and therefore scatter radiation very poorly. As a result, when all the electrons in the Universe suddenly jumped aboard atoms, the Universe became transparent.

Fossil radiation

With nothing to bump into any more, the radiation generated 500,000 years after the Big Bang has survived until today, almost 15 billion years later, virtually unchanged. Well, almost. The Universe has been constantly expanding throughout that time, and the expansion has stretched out the radiation's wavelength. At 500,000 years, most of the radiation had a wavelength of one-millionth of a millimetre, corresponding to ultraviolet light. Fourteen and a half billion years of cosmic expansion has stretched this out to about one millimetre, corresponding to microwaves. At this wavelength, the radiation has a characteristic temperature of -270°C – only 3° above absolute zero (which is as cold as it is possible to get). And these supercooled microwaves are known as cosmic microwave background radiation (CMBR).

That the CMBR is a consequence of the Big Bang theory was first realized by Russian-US physicist George Gamow, together with his students Ralph Alpher and Robert Herman, in 1948. Their crude

1. George Gamow led the team who first calculated how the Big Bang fireball behaved.

There are some 50 billion galaxies visible from Earth; the nearest to us is Andromeda, 3 million light years away.

calculations showed that if the Big Bang theory was correct, then the Universe today should be bathed in radiation at about 10° above absolute zero, which is roughly right. If this radiation could be detected, it would be a big piece of evidence in favour of the Big Bang – supporting the idea that the Universe has been expanding at least since it was 500,000 years old, and that its temperature at that time was consistent with the predictions of early-universe physics.

In fact, an indirect detection of the CMBR had already been made during the 1930s. Astronomical studies had shown that clouds of interstellar gas in the Milky Way do indeed have temperatures of around -270°C. Surprisingly though, this failed to suggest to Gamow and his team that they should try to lower their estimates, and the trail went cold.

Expect the unexpected

It was not until the 1960s that new teams of researchers approached the problem afresh – not all of them intentionally. Arno Penzias and Robert Wilson of Bell Laboratories, New Jersey, were adapting a radio antenna, which had originally been developed for satellite communications, for use in radio astronomy research. They had no particular inclination to use the antenna to search for the CMBR, but they did intend to use it for studying astronomical radio sources that were extremely weak. This meant that the antenna had to be very sensitive and that every possible source of radio noise had to be identified and eliminated.

But one source of noise refused to go away. Penzias and Wilson's equipment was picking up a faint crackle of radiation with a temperature of -270°C. None of the precautions they took –

1

1. The bright band of the Milky Way is the central disc of the spiral galaxy that our Sun is part of. The Sun's motion around the galaxy blueshifts and redshifts the CMBR.

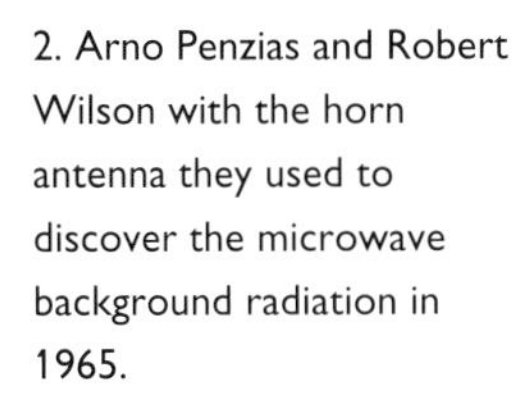

2. Arno Penzias and Robert Wilson with the horn antenna they used to discover the microwave background radiation in 1965.

2

including the removal of large amounts of pigeon droppings from the antenna horn – could eliminate the noise.

At about the same time, P.J.E. Peebles, an astrophysicist at Princeton University, had independently reproduced the earlier calculations of Gamow and his colleagues – showing that space should be pervaded by a relic microwave glow. Penzias and Wilson soon learned about the work of Peebles, and contacted his supervisor, Robert Dicke, in the hope that his findings could possibly explain the noise plaguing their antenna. Indeed they could. Penzias and Wilson were, of course, seeing the cosmic microwave background. The two groups each published separate research papers setting out their respective parts in the discovery, which appeared side by side in the *Astrophysical Journal* in 1965. In 1978, Penzias and Wilson rightly received the Nobel Prize for their role in what is the single most significant piece of evidence there is for the Big Bang theory of the Universe.

Lumpy and bumpy

When Penzias and Wilson first identified the CMBR, it appeared extremely smooth, having the same temperature right across the sky. However, as detector technology improved, small variations in the radiation eventually became visible.

The first variation was the so-called CMBR dipole. This is caused by the motion of the Earth through space (due to a combination of the motion of the Milky Way galaxy through space, the Solar System's orbit around the Milky Way and the Earth's orbit around the Sun). The overall effect is to make the background radiation in the direction

WRINKLES IN TIME

In 1992, NASA's Cosmic Background Explorer spacecraft (COBE, below) stunned the scientific world when it detected the seeds in the young Universe from which galaxies – and ultimately stars, planets and people – grew. COBE was an Earth-orbiting satellite, armed with sensors to measure the microwave background radiation, the dilute electromagnetic echo left over from the Big Bang fireball. The sensors picked out minuscule lumps and bumps that had been imprinted on the microwave background by irregularities in the density of matter from point to point throughout space. It was these irregularities, generated in the first instants after the Big Bang, that grew by gravity into the vast cosmic superstructure of galaxies and clusters that occupies the Universe today.

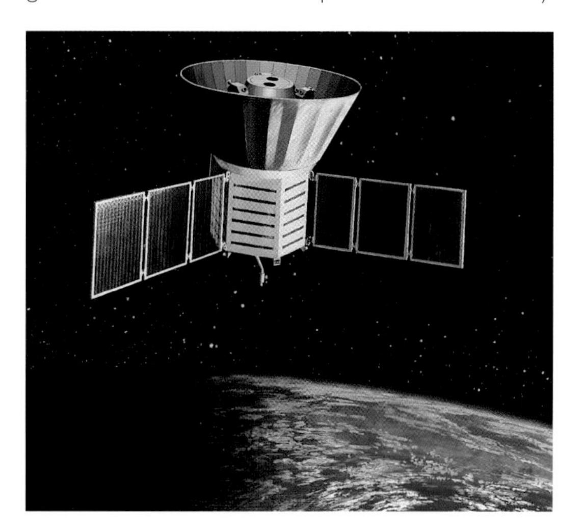

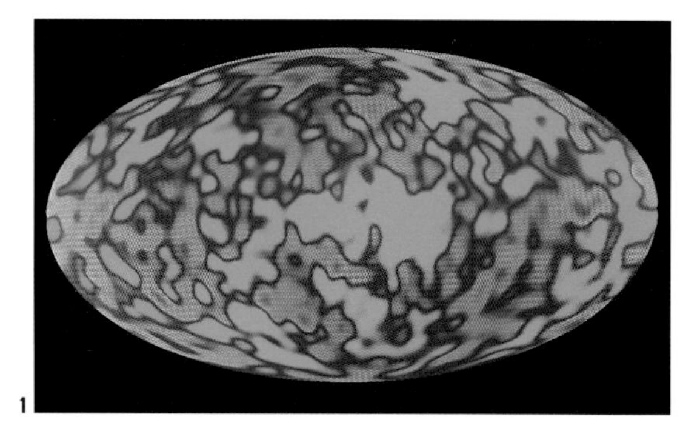

1. Microwave map of the night sky from COBE. Ripples on the CMBR were blown up to astrophysical scales by cosmic inflation.

of the Earth's net forward motion appear slightly hotter than the radiation behind it. This is due to the Doppler effect blueshifting the part of the sky that the Earth is moving towards and redshifting the part that it's moving away from (▷ p. 22). The dipole was first observed in the late 1970s by detectors placed aboard high-altitude balloons and aircraft.

Much smaller variations in the CMBR have since been seen. These are the irregularities from which the galaxies are thought to have grown. They were first picked up in 1992, by the COBE satellite, and recently studied in greater detail by the Boomerang balloon experiment flown over the Antarctic.

In 2001, NASA will launch its Microwave Anisotropy Probe (MAP), which will make even more detailed studies of the variations in the CMBR from point to point across the sky. This will enable astrophysicists to perfect their theories of how the galaxies formed.

THE ORIGIN OF THE LIGHT ELEMENTS

Observations of the CMBR tell cosmologists that the Big Bang theory holds good for the Universe back to when it was about 500,000 years old – a tiny fraction of its present age. But there is another crucial observation that extends confidence in the theory back much further, to one-hundredth of a second after the moment that the Universe was created, when the temperature was a searing 100 billion°C.

At this time, particle physics processes were transmuting electrically neutral subatomic particles called neutrons into similar but positively charged particles called protons, and back again. Protons are slightly lighter than neutrons, meaning that they have a slightly lower energy. As the Universe continued to expand and cool, and its energy density therefore diminished, the lower energy state of the protons became preferable to the higher energy of the neutrons. An excess of protons therefore began to accumulate.

The transmutation process involved interactions with a species of tiny particles called neutrinos. But, at one second after the Big Bang, the temperature dropped to 10 billion°C – too low for neutrinos to interact with other particles. Numbers of protons and neutrons then suddenly became fixed, at a ratio of about seven protons to every one neutron. As the nucleus of hydrogen – the lightest chemical element – is made of just a single proton, the Universe at this time was essentially a sea of hydrogen nuclei.

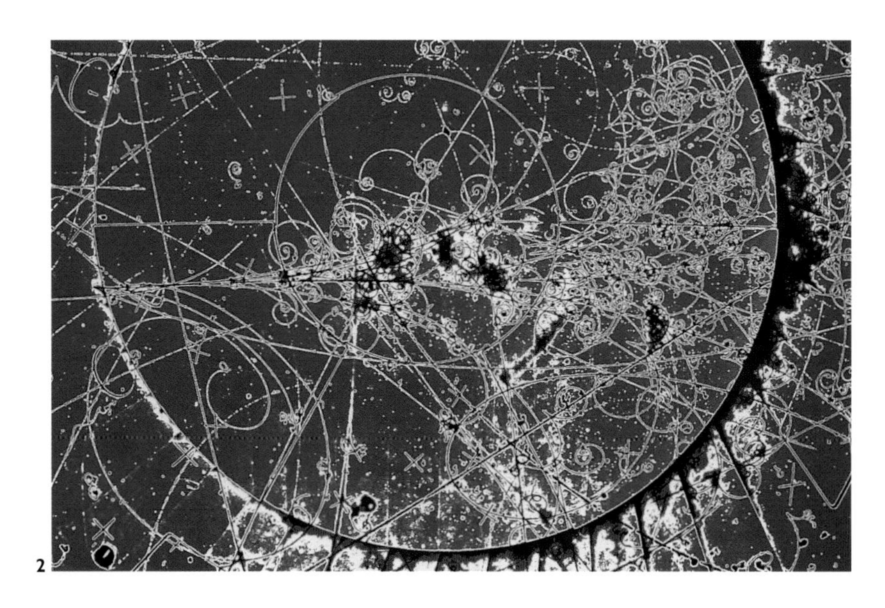

2. A mass of entangled subatomic particle tracks, as recorded in a bubble chamber at the particle accelerator laboratory at CERN. Similar processes took place very shortly after the Big Bang, and probably gave rise to the balance of chemical elements in the Universe today.

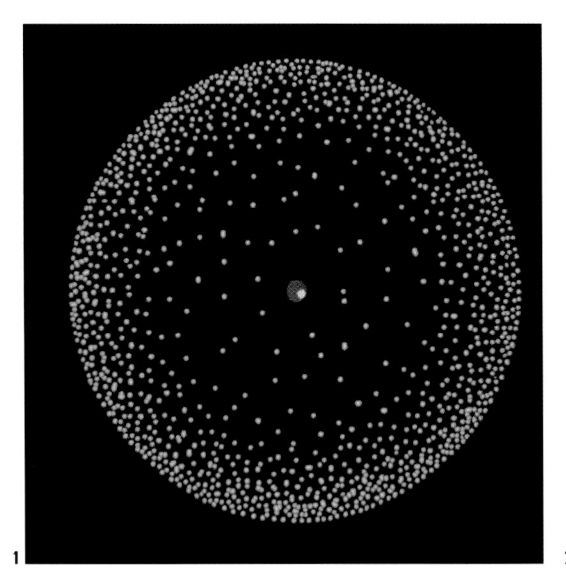

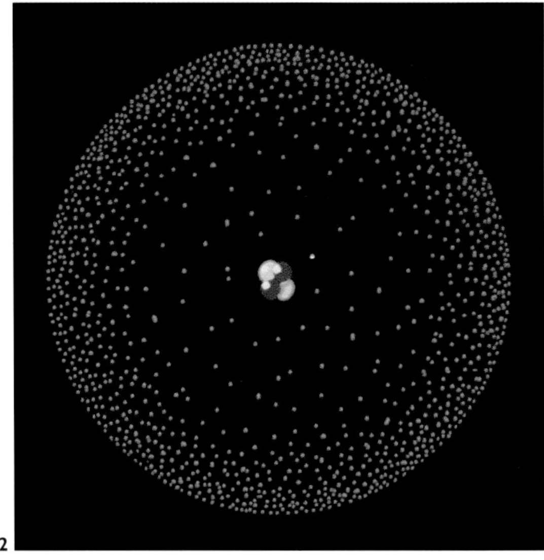

1. How a hydrogen atom might look close up: a single proton (red) is surrounded by an electron cloud (blue).

2. Computer graphic of a helium atom: a nucleus consisting of two protons (red) and two neutrons (blue) in an electron cloud.

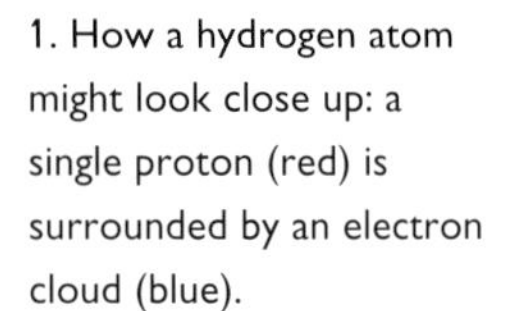

If the Universe had expanded more slowly during the formation of atomic nuclei, it would be made almost entirely of iron – stars and life would not have emerged.

By the time the Universe had reached an age of 15 seconds, the temperature was down to 3 billion°C. That's cool enough for a proton to pick up a neutron and form a nucleus of so-called 'heavy' hydrogen, also known as deuterium. Soon after, at about three minutes (temperature now 1 billion°C), these deuterium nuclei started to pair up to form nuclei of helium (comprising two neutrons and two protons).

Since the Universe at this time contained seven protons for every one neutron, six out of every seven protons were left partnerless, destined to remain as hydrogen. Protons and neutrons are by far the heaviest particles in nature, so these partnerless protons made up about 75 per cent of the Universe. The other 25 per cent – one proton out of every seven, plus a neutron – became deuterium, which soon merged to become helium.

And this is the material content of the Universe as it emerged from the Big Bang – 75 per cent hydrogen, 25 per cent helium. At least that's the theory.

Cosmic abundances

To verify these numbers, astronomers have looked at the light from old stars, formed very early in the history of the Universe, and examined each star's spectrum, obtained by splitting up the light into individual colours and measuring the brightness of each colour. The positions of peaks and dips in the spectrum reveal what elements are present inside a star. And the respective height or depth of each peak or dip shows how much of each element there is. Astronomers find that old stars have exactly the 75 per cent hydrogen/25 per cent helium composition predicted by the Big Bang theory.

Stars are effectively giant nuclear reactors, which process the light chemical elements, formed during the early moments of the Universe, into heavier, more complicated ones. New stars form from the remains of earlier generations. Each new generation has a slightly different initial composition from the previous one, and therefore presents an increasingly inaccurate picture of the abundances of elements

3. Omega Centauri (NGC 5139) is one of the largest and brightest globular star clusters visible in the night sky. Globular clusters contain some of the oldest stars in the Universe, and as such are excellent for studying the chemical elements produced in the Big Bang.

formed in the Big Bang. This is why old stars are essential for studies of the primordial abundances.

The nuclear processing power of stars produced most of the chemical elements heavier than helium in the Universe. These elements were forged in the centres of massive stars and then flung across space when the stars blew themselves apart in supernova explosions. Look around you. Anything that is not hydrogen or helium (or lithium, a very small portion of which was also formed in the Big Bang) was made long ago inside a star. That includes the carbon in your body and the oxygen you breathe.

The usual suspects

The calculations needed to work out what elements were formed in the hot fires of the very early Universe were first performed by George Gamow, Ralph Alpher and Robert Herman in the 1940s. These same people made the initial prediction of the cosmic microwave background radiation. Before this, in the late 1920s, Gamow had also used quantum physics to build a theory of radioactivity – a phenomenon that can be turned on its head to explain the very same nuclear fusion reactions that were responsible for forging the light elements in the Big Bang.

Gamow's calculations in the 1940s were refined in the 1960s by British astronomer Sir Fred Hoyle and, independently, by Russian astrophysicist Yaakov Borisovich Zel'dovich.

Explaining the synthesis of the light chemical elements is one of the great successes of the Big Bang theory. But it would turn out that not everything in the Universe is made from simple chemicals. By the time creation of the light

ALPHA BETA GAMMA

In the 1940s, cosmologist George Gamow and his student Ralph Alpher used the laws of nuclear physics to calculate what chemical elements would have been forged in the hot furnace of the Big Bang. Their results have been confirmed by astronomical studies of the elements found in very old stars. While writing up their findings for the journal *Physical Review*, Gamow noticed the similarity between the authors' names and the first and third letters of the Greek alphabet – alpha and gamma. Feeling it unfair to omit the second Greek letter, beta, Gamow added a credit to German-US physicist Hans Bethe (right) between his name and Alpher's – even though Bethe had not contributed to the work. Appropriately, the paper appeared in the 1 April 1948 edition of the journal. Its contents are still referred to as the 'alpha beta gamma' theory.

1. The Cygnus Loop, 2500 light years away in the constellation of Cygnus, is the remnant from a supernova explosion that occurred 15,000 years ago. Supernovas are responsible for seeding the Universe with the chemical elements heavier than helium.

elements in the Big Bang was understood, astronomers had already begun to notice that there didn't seem to be enough normal, bright matter in the Universe to explain the gravitational motions of the stars and the galaxies. In fact, it seemed as though 95 per cent of the Universe was missing. If gravity was to be believed, there is something else occupying the Universe – something dark.

DARK MATTER

Looking up at the sky on a clear night reveals a vast abundance of stars and galaxies. But this represents only a small fraction of the entire mass of the Universe. Astrophysicists know this because celestial objects are moved around by the action of gravity, which is generated by matter. Explaining the observed gravitational motions of stars and galaxies requires about 20 times more matter than can be seen. This ongoing conundrum is known as the missing mass, or dark matter problem.

The dark cosmos

The evidence for missing mass is extremely wide-ranging. The first clue emerged in the 1920s, when Dutch astronomer Jan Oort measured the speeds of stars moving up and down through the disc of our Milky Way galaxy. The gravity of the disc makes the stars bob up and down like carousel horses as they orbit the galaxy's centre. But Oort found that the stars were not moving as far up or down on each 'bob' as they should. He concluded that there must be about 50 per cent more matter in the Milky Way's disc than could be seen, and that the gravity of this matter was restricting the stars' motions.

Galaxies cluster on to giant pancake-like structures, separated by enormous voids. Voids occupy some 98 per cent of the Universe.

1

1. Vera Rubin (right) and colleagues found key evidence for dark matter's existence in spiral galaxies.

2. (opposite) Starfield from a densely populated area of the Milky Way galaxy, centred on the constellation of Musca.

In the 1970s and 1980s, American astronomer Vera Rubin and her colleagues found more evidence in other spiral galaxies. Rubin's team studied how the orbital speed of stars around spiral galaxies varies with the distance from each galaxy's centre. Based on calculations of the gravitational field in galaxy discs, Rubin's team expected the stars' speeds to decrease steadily with distance. But instead they found that the orbital speed was roughly constant. This could mean only one thing: each galaxy is surrounded by a halo of invisible material, weighing 10 times as much as the bright

1

1. Galaxy cluster Abell 2218, seen from the Hubble Space Telescope. To provide enough gravity to prevent them from flying apart, galaxy clusters must harbour more mass than can be seen. Their dark matter is thought to weigh about 20 times their visible matter.

disc. The stars' speeds are held constant by the gravity of this invisible halo.

Galaxies are often grouped into clusters. And these clusters have presented yet more evidence for dark matter. Studies in the 1980s showed that many galaxies in clusters appear to be moving so fast that the clusters should be flying apart – not enough gravity is generated by their visible matter to hold them together. The fact that they are clearly not flying apart means that the extra mass needed must be there, but in some invisible form. Calculations show that the invisible matter must weigh about 20 times the mass of the cluster's bright component.

So what is it?

Most astrophysicists believe that the dark matter of the Universe probably takes the form of exotic subatomic particles. These particles come in two possible types: hot dark matter and cold dark matter. Hot dark matter comprises a huge number of very light particles, known as neutrinos. These move very rapidly – close to the speed of light – and therefore carry a lot of energy, hence the name 'hot'. Initially, neutrinos were thought to weigh nothing. But new evidence suggests that they do have a very small mass. Multiply this tiny mass by their vast numbers (countless billions pass through

your body every second) and you have one possible explanation for the missing mass of the Universe.

Cold dark matter, on the other hand, would be made up of a smaller number of heavier particles, which move more slowly and so have lower energy – hence 'cold'. Cold dark matter particles have yet to be detected experimentally, but are predicted by supersymmetry – the group name given to

1. Galaxy cluster NGC 2300 seen from the orbiting ROSAT X-ray observatory. The galaxies are embedded in a bright cloud of gas, which is emitting X-rays (in magenta).

currently fashionable theories that unify the different families of subatomic particles.

It's also possible that some of the missing mass is accounted for by 'dark energy' – energy locked away in the fabric of empty space. As Einstein showed with his equation $E = mc^2$, mass and energy are equivalent, so dark energy creates gravity just as dark matter does. The key difference, however, is that dark energy, if it exists, should make the expansion of the Universe accelerate on large scales. In 1998, scientists led by Saul Perlmutter of Lawrence Berkeley National Laboratory, California, made accurate measurements of the expansion rate of the Universe. They found that it does seem to be accelerating, indicating that 70 per cent of the mass of the Universe could be composed of dark energy.

Is dark matter dead?

The most recent development in the story of dark matter is the finding that the missing mass of the Universe might not be missing after all. That's the conclusion of Stacy McGaugh of the University of Maryland. He has analysed the results from Boomerang, a balloon-mounted experiment flown over the Antarctic,which studied the size of the irregularities in the CMBR in unprecedented detail.

Astrophysicists analyse the size of CMBR irregularities by plotting a graph, known as a power spectrum, showing how many irregularities there are of each particular size. Dark matter theories predict the existence of a peak in the power spectrum, the height of which is linked directly to the amount of dark matter there is in the Universe.

DETECTING DARK MATTER

British scientists have set up an experiment in England to try to detect particles of so-called cold dark matter, which could make up a large fraction of the Universe's mass. The experiment, down a potash mine near Boulby, North Yorkshire, consists of a light detector – a 6-kg (13-lb) sodium iodide crystal – with photomultipliers mounted either side of it (right). The scientists hope that every so often a dark matter particle will collide with an atomic nucleus in the crystal and release some of its energy as a faint pulse of light, which can then be picked up by the photomultipliers. The detector is placed under ground because the surrounding rock acts as an excellent screen to block out cosmic radiation in the Earth's atmosphere, which could contaminate the results. The detector is also immersed in a 200-cubic-metre (7060-cubic-ft) tank of water to block out neutron and gamma radiation from the rock.

1. If dark matter is composed of many weakly interacting massive particles (WIMPs) streaming through space, then the 'empty' space between stars is full of subatomic particles.

2. Map of the CMBR from the Boomerang balloon-mounted experiment flown over the Antarctic in 1998–9. One interpretation of the Boomerang results is that dark matter may not exist after all.

But when McGaugh plotted the Boomerang observations, he found that this peak was missing.

McGaugh concludes that dark matter doesn't exist. Instead, he thinks Boomerang's findings warrant a modification of the laws of gravity. McGaugh finds that the Boomerang data match exactly with the predictions of so-called modified Newtonian dynamics (MOND) – an alternative gravity theory put forward in the early 1980s. Einstein's general relativity is our current best description of gravity. But in weak gravitational fields, such as those governing the behaviour of big diffuse galaxies, Newton's 300-year-old gravity theory is a good enough approximation. MOND represents a modification of general relativity in this so-called weak, or Newtonian, limit.

The claim will be put to the test in 2001, when NASA's Microwave Anisotropy Probe (MAP) spacecraft will study the CMBR in more detail than ever. It could finally solve the dark matter problem.

SMOOTH AND FLAT, BUT WHY?

We still don't know what 95 per cent of the Universe is made from. The dark matter mystery is probably the single most serious problem with the Big Bang theory. Two others – the so-called horizon and flatness problems – were solved by the idea of inflation – an extension to the Big Bang theory which says that the Universe had a growth spurt during its earliest moments (▷ Chapter 2).

Cloud on the horizon

The horizon problem can be summarized by asking the question of why opposite sides of the night sky look the same. True, there are differences – different constellations on each side of the sky, different galaxies and different clusters of galaxies. But there are no radical differences. For instance, we do not see one side of the sky blazing brightly while the other appears completely dark. To a casual observer the night sky looks more or less the same in every direction.

But why should it? The Universe has been around for some 15 billion years, so the most distant objects that we can see are 15 billion light years away. Opposite sides of the visible night sky are then 30 billion light years apart. Because nothing can travel faster than the speed of light, it is impossible for any signals or natural processes to have reached from one side of the sky to the other. So why should they look the same?

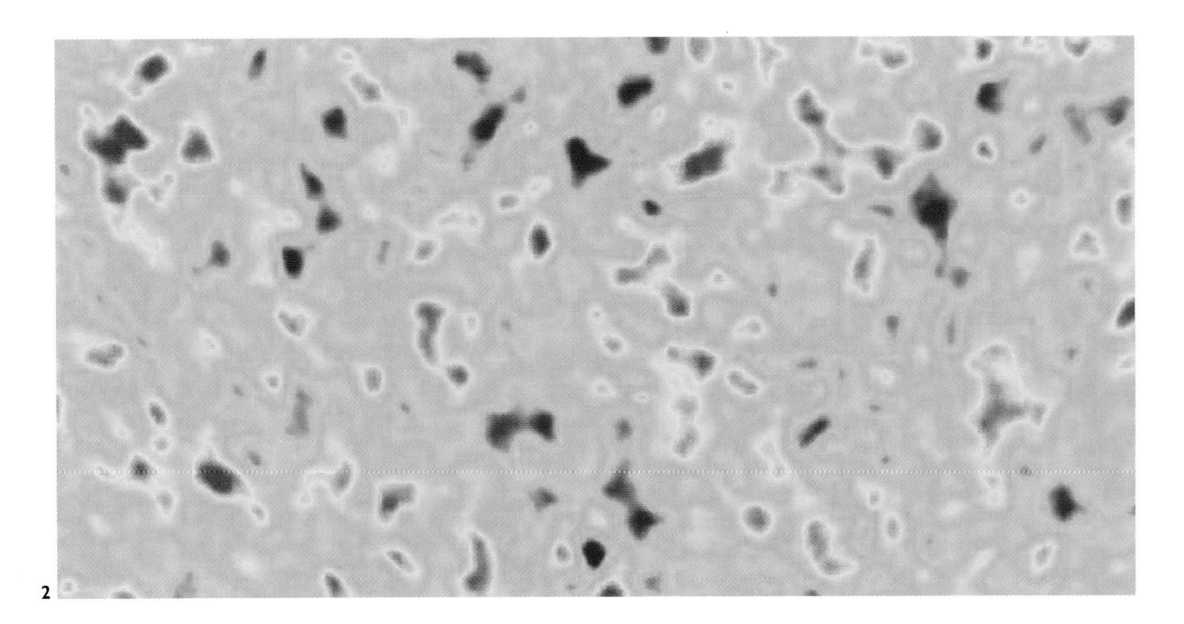

2

Think of it another way. If you quickly empty a pot of treacle on to one side of a plate, the treacle is initially concentrated in one big dollop, so that side of the plate looks different from the other. In a few minutes, the treacle spreads out to cover the whole plate evenly. Similarly, the chaos of the Big Bang should have produced some regions of the Universe that looked very different from others – the cosmic equivalent to dollops of treacle. But in this case, the dollops take 30 billion years to smooth out. If the Universe is only 15 billion years old, why does it look so smooth? This is the horizon problem.

Inflation solves the problem by rapidly expanding the very early Universe. It means that what we see as the entire visible Universe today was inflated from a tiny piece of the early Universe. This initial piece was so small that, before inflation, natural processes could easily have travelled from one side of it to the other, thus making it smooth.

Going back to the treacle analogy, inflation means that rather than likening the visible Universe to the entire plate, we should think of it as just a small region on the plate, say the size of a small coin. If you zoom in on a coin-sized region

1. The Universe as it was almost 15 billion years ago. This 'deep field' image from the Hubble Space Telescope spans just one-hundredth of the diameter of the full Moon, and was assembled from over 200 exposures.

2. If today's Universe is roughly flat, the early Universe must have been almost exactly flat, to an astonishing degree of accuracy. Cosmological inflation explains why the early Universe was so flat.

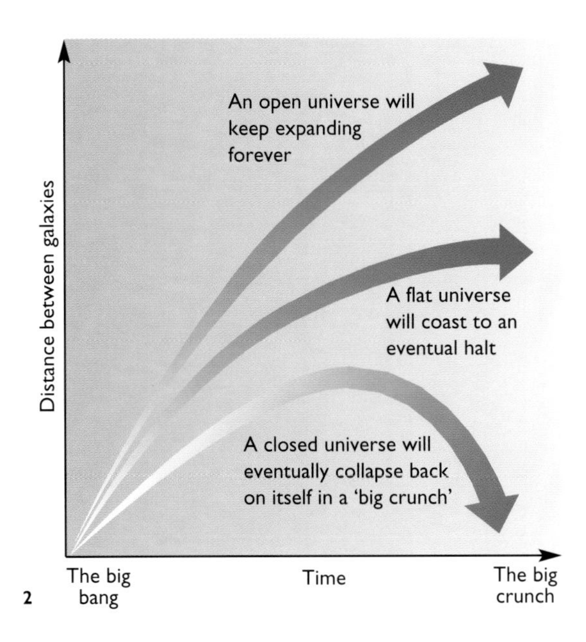

on a dollop of treacle, it looks quite smooth. And that's how inflation solves the horizon problem.

The flatness problem

Einstein's general theory of relativity dictates how space is curved by the matter and energy that it contains. Applied to cosmology, it predicts that the Universe at large can take one of three possible shapes, known as closed, open and flat. A closed universe would be shaped like a sphere – wrapped around, or 'closed' in on, itself. An open universe, on the other hand, would have a shape rather like a saddle, but extending to infinity in all directions. On the knife edge between these two possibilities is a flat universe. In this case, the gravitational and mass energies of the Universe cancel out, and space on the largest scales has zero curvature.

As curvature is determined by matter, cosmologists often classify the shape of the Universe by a number that determines the density of the matter that it contains. This number is called the density parameter and is denoted by the Greek letter omega (Ω). If $\Omega = 1$, the Universe is flat; if it's less than 1, the Universe is open; and if it's greater than 1, then the Universe is closed.

Astronomers today are reasonably certain that Ω is 1 to within an order of magnitude in each direction. That might not seem too constraining. However, $\Omega = 1$ is what mathematicians call a 'repeller'. That means that any small deviation from $\Omega = 1$ becomes amplified rapidly as the Universe grows older. So much so that for Ω to be within a stone's throw of 1 today means that when the Universe was one second old, it must have been almost exactly 1, to within 1 part in 10^{60}. But the problem is, why was the early Universe so flat?

Inflation solves the flatness problem very neatly. Rapidly inflating the Universe makes it very much larger in a very short time. Imagine that the Universe before inflation looked like a beach-ball. It's plain to see when you hold the ball in front of you that its surface is curved. But if you inflate the ball to the size of the Earth, the curvature is imperceptible and it looks flat. That's what cosmological inflation did to the Universe. In fact, inflation expanded the early Universe so much that expanding a beach-ball by the same factor would produce a ball 10^{40} times the size of the visible Universe – and that would look convincingly flat.

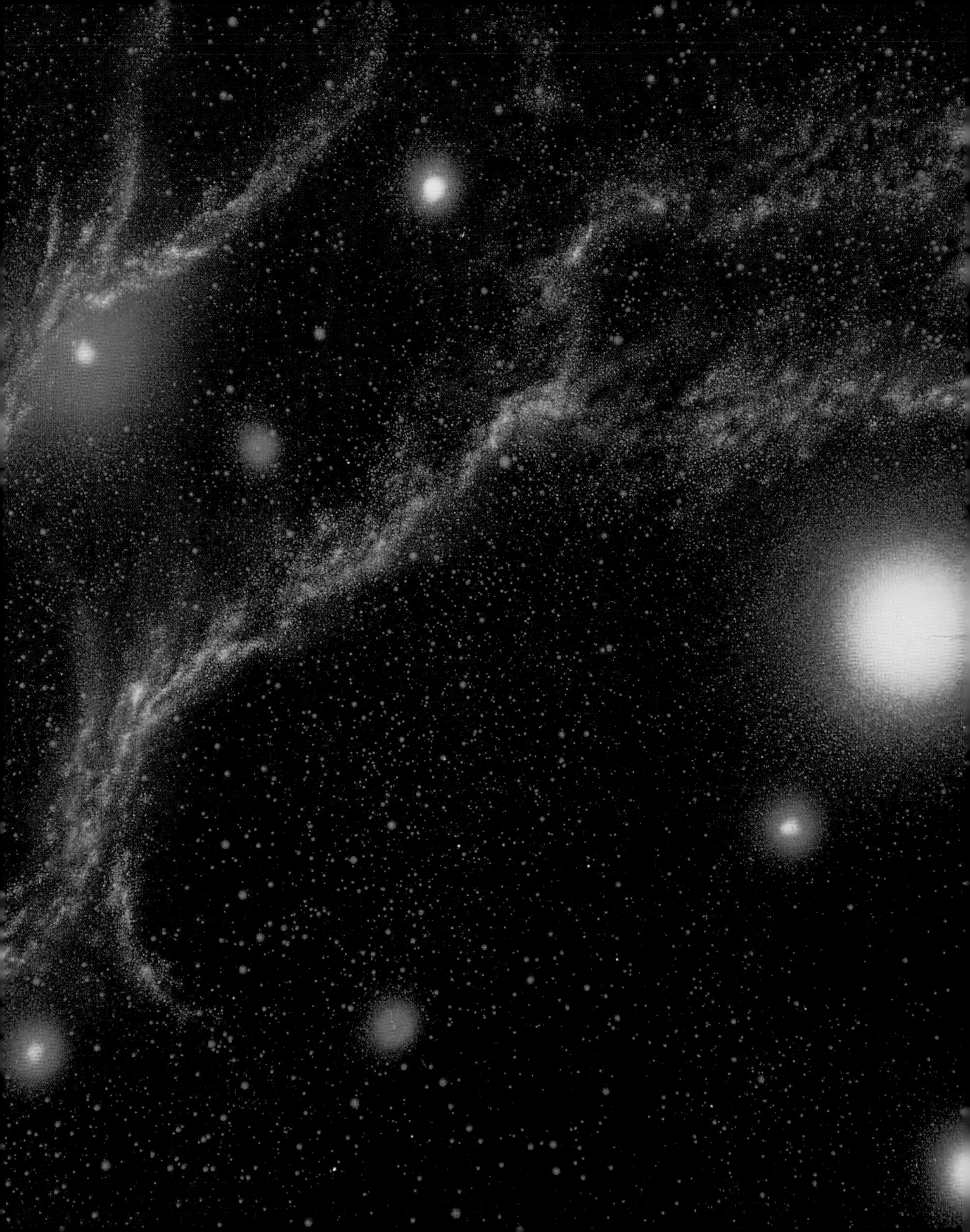

4

INTO THE UNKNOWN

4

INTO THE UNKNOWN

What does the future hold for our Universe? Will the cosmos re-collapse on itself, burning out in what scientists call a 'Big Crunch'? Or will it fade away, expanding for ever to become a black, featureless void? And how will our understanding of the Universe change over the years to come? If the history of science has taught us anything, it is that there are probably countless discoveries about the Universe still waiting to be made. Perhaps one day we will unify the fundamental forces of nature, discover the truth about the Universe's elusive dark matter, or find out how to burrow into higher dimensions of space and time. Who knows? For the moment we can only speculate. Some physicists have even supposed that our Universe might be the tip of a cosmic iceberg comprising many universes, linked into a grand transuniversal network known as the multiverse. The multiverse could also explain many of the mysteries of quantum physics.

Previous page: Dying giants. Some scientists say that, in time, the stars will begin to burn out and die, matter will decay away, and life as we know it may no longer be possible. The death of the Universe will have begun.

CRUNCH TIME?

In the *Hitch Hiker's Guide to the Galaxy* novels by Douglas Adams, the restaurant at the end of the Universe was a place (or, perhaps more accurately, a time) where cosmic travellers could relax over a meal and watch the death of the Universe spectacularly unfold.

Although *Hitch Hiker's Guide* was a comedy, the death of the Universe is a depressing thought. This is the end of everything. Not just one planet, one star, or one galaxy, but absolutely everything that we gaze up at in awe on a clear evening. Somehow it is poor consolation that we'll all be long dead by the time it happens. So just how will it happen?

Cause of death

Cosmologists have come up with two possible scenarios for the ultimate fate of the Universe. Which one is correct depends to a large extent on the value of one number, introduced in Chapter 3, called the cosmic density parameter. Denoted by the Greek letter omega (Ω), the density parameter is basically a measure of how much material there is in the Universe and so, through the laws of gravitation, how much gravity there is. If Ω is less than or equal to 1, then there is not enough gravity to stop space expanding, and the Universe will keep on forever getting bigger.

However, if Ω is greater than 1, the Universe will one day stop expanding and start to contract. A contracting universe is destined to collapse back to a point of infinite temperature and density, similar to the conditions that prevailed in the Big Bang. In this scenario, Armageddon quietly begins. The first thing that astronomers of the future would notice is that the redshifts of distant galaxies don't seem to be as big as they used to be – the motion of the galaxies away from us appears to be slowing down. As expansion gradually turns into contraction, redshifts would become blueshifts; galaxies would

1. Cosmic expansion makes distant galaxies appear redder than the nearer ones. In the Big Crunch scenario, this cosmic redshifting would lessen and then turn into a blueshift.

stop getting farther apart and start getting close together again.

The history of the Universe that cosmologists have struggled to piece together throughout the 20th century would now be played out in reverse. The cosmic microwave background – heat from the Big Bang, supercooled by billions of years of cosmic expansion – would start to reheat. Eventually, it becomes hotter than stars, making the stars themselves break apart. Galaxies crash together and overlap. Matter is swallowed up by the giant black holes in the centres of galaxies. The particle physics processes that brought the forces of nature into existence during the Big Bang are undone, and physics becomes governed once more by a single unified superforce. And then, just as quickly as it was born, the Universe is gone.

It's called the Big Crunch, but this cataclysmic antithesis to the Big Bang does have a final optimistic twist. Many cosmologists believe that were a Big Crunch actually to happen, rather than being snuffed out of existence, the collapsing Universe might 'bounce back', rising phoenix-like from the ashes to begin a new phase of cosmic expansion. This could be how our own Universe began. But it is unlikely that we will ever know for certain.

Burn out or fade away?

Despite its pleasing symmetry, the Big Crunch scenario seems unlikely. Many cosmologists believe that Ω is in fact less than 1, meaning that the Universe will expand for ever. Others say that space is ever-expanding for a different reason. Observations in the late 1990s by an international

THE GÖDEL UNIVERSE

One of the weirdest alternatives to the Big Bang expanding universe theory was conceived in the late 1940s by Austrian-US mathematician Kurt Gödel (near right, with Albert Einstein). Gödel wondered if the tendency of gravity to make the expansion of the Universe slow down, and even re-collapse, could be countered by making the Universe rotate so that outward centrifugal forces balanced gravity's inward pull. The result – the Gödel universe – appears to rotate about the observer, wherever the observer is situated. Stranger still, Gödel found that anyone travelling around a large enough loop in his universe can travel backwards through time – but the time travel loops would need to be about 100 billion light years long. Astronomical observations indicate that it's highly unlikely we live in a Gödel universe.

1. How the Universe might end in the Big Crunch: stars, galaxies and dust clouds are drawn into a black hole in a brilliant spiral.

team of astronomers, led by Saul Perlmutter of Lawrence Berkeley National Laboratory, California, have shown that space seems to be pervaded by what is called 'dark energy'. This energy, which could account for as much as 70 per cent of the mass of the Universe, has the unusual effect of making the expansion of space accelerate.

The idea of dark energy was first introduced soon after the discovery of general relativity. Referred to as the 'cosmological constant', it was initially brought into play, ironically, to stop theoretical models of the Universe from expanding. This was before Hubble and Humason had discovered that cosmic expansion was a real effect. Dark energy enters modern-day theories with the opposite mathematical sign to the original cosmological constant, so that rather than halting the expansion, the energy accelerates it – making the Universe expand for ever.

If space is packed with dark energy – or if the cosmic density parameter, Ω, is less than or equal to 1 – we are saved from the Big Crunch. But little good it will do us. If the Universe eternally expands,

Albert Einstein first introduced the cosmological constant in 1917, and later removed it again, calling it his 'biggest blunder'.

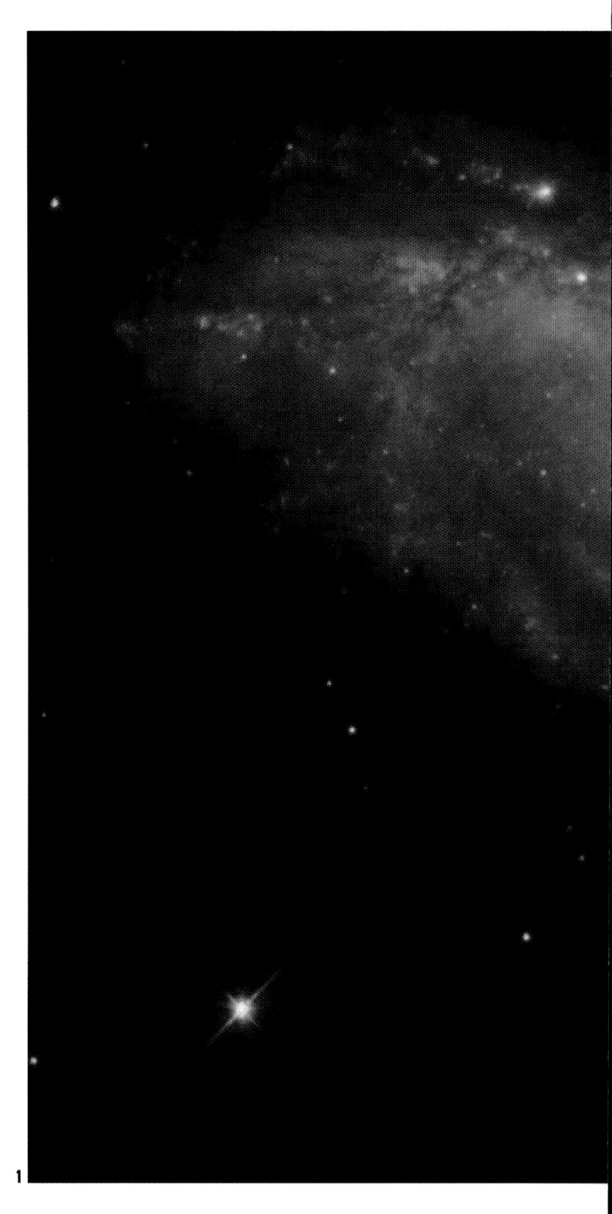
1

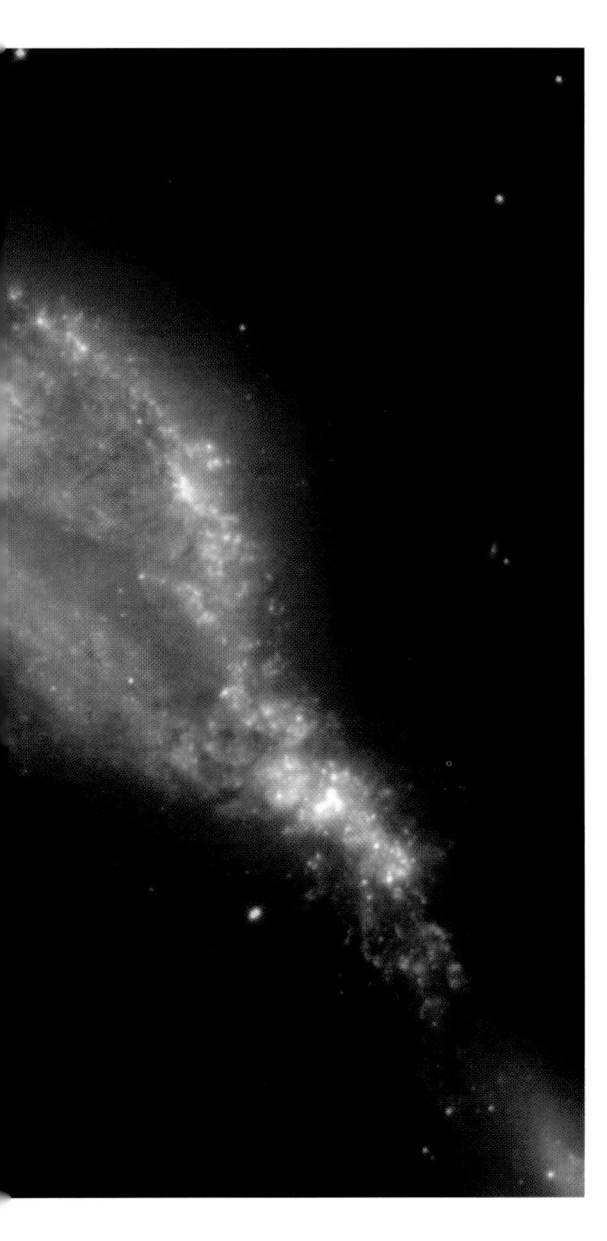

1. As the Big Crunch ensues and space collapses on itself, galaxies would move closer together and eventually collide. Shown here is a galaxy collision observed by the orbiting Hubble Space Telescope.

the density of matter and the radiation within it are destined eventually to dwindle away to zero. The stars will burn out and die. The galaxies will gradually switch off. And ultimately even fundamental particles of matter will decay away, leaving just space in all its vast, empty blackness. This scenario for the fate of the Universe is known to cosmologists as the heat death.

Last entertainment

At present the Universe is in the so-called stelliferous era – the age of stars. In the heat death scenario, the beginning of the end for this grand cosmic age will happen in about a trillion (1000 billion) years' time. At that point, raw star-forming materials – hydrogen and helium gas – will have become so depleted that there is nothing left from which to build new generations of stars. Soon after, the last geriatrics in the last generation of stars die, and the Universe then enters a new and bleak cosmic age known as the degenerate era.

Nothing to do with cosmic moral standards, degeneracy here refers to a very dense state of matter, in which subatomic particles are packed together extremely tightly. The Universe during the degenerate era is composed largely of

1

1. The Eagle Nebula is a star-forming region, rich in gas and dust – the raw materials for new stars. In the heat-death scenario for the end of the Universe these raw materials will be used up, and stars will become a thing of the past.

2. An artist's impression of the Big Bang. Understanding this earliest instant in cosmic history will require us to unify the laws of quantum physics and gravity.

superdense white dwarfs and neutron stars – stellar corpses that are made almost entirely from degenerate material.

But even this is not immortal. By the time the Universe celebrates its trillion-trillion-trillionth birthday, white dwarfs and neutron stars will also have gone. They – as well as most of the dark matter in the Universe and any remaining gas, dust and planets – will have been eaten up by the black holes that today lurk at the centres of most galaxies. Any protons or neutrons left will have decayed into radiation and other particles, which in turn are also swallowed up by the black holes. Indeed, these cosmic garbage collectors are now all that is left; the Universe has entered the black hole era.

This cosmic age lasts for a mind-boggling 10^{100} years – that's 1 with 100 zeros after it. By comparison, the Universe we live in today is a sprightly 10^{10} years old. At the end of the black hole era, black holes themselves evaporate away in a puff of particles and radiation (a process that was discovered by physicist Stephen Hawking in the early 1970s). Finally, as the Universe continues to expand and cool, the concentration of these particles and radiation becomes steadily diluted away to zero. At that point space is truly empty, and the Universe is dead.

THE GROWTH OF KNOWLEDGE

A few scientists believe that science is nearing its end. We are running out of things to discover, they say. Yet many believed that science was more or less complete 100 years ago, before quantum theory and relativity – two great cornerstones of science, without which it is virtually impossible to imagine modern physics today – were even formulated. The chances are that there's still plenty for us to discover. And if history is anything to go by, this will be especially true of the cosmos at large.

The holy grail of modern cosmology, not to mention particle physics, is to find the underlying theory that unifies the four forces of nature – gravity, electromagnetism and the strong and weak nuclear forces. Today these forces exist as distinct entities. But during the earliest moments after the Big Bang they were entwined together into a single superforce. It was this superforce that governed the behaviour of the Universe when the fires of the Big Bang were at their hottest. And that is why it's so important that we get to grips with it.

2

As the young Universe expanded and cooled, the individual forces peeled away from the superforce one by one – gravity went first, then the strong force, followed by the weak force. How electromagnetism and the weak nuclear force fit back together was discovered in the late 1960s – the so-called electroweak theory – and is now well established. Scientists also have a rough idea of how the strong force might fit into the picture. The real bugbear, however, is gravity.

Matters of gravity

It is gravity that determines how the Universe expands. However, tracing cosmic expansion backwards in time we arrive at a point when the Universe was extremely small, and so must also have been governed by the laws of the very small: quantum physics. The implication is that in order to understand the very early Universe, we must somehow make the laws of gravity and quantum physics mesh together. The trouble is that the quantum laws seem utterly irreconcilable with the best theory of gravity that we have today – Einstein's general theory of relativity. But why?

Quantum theory can be something of a black art. Physicists make sense of the results of quantum calculations by using a technique called 'renormalization'. Sometimes the equations of quantum theory yield infinite answers – numbers so large that they can't be real. The renormalization procedure gets rid of these unwanted infinities and keeps only the more manageable, finite answers. It's a bit of a fudge, but it's a consistent fudge – and it works. Physicists are therefore happy to use it – at least until they can think of something better.

However, try as they might, they can't make renormalization remove the infinities that crop up in theories of quantum gravity which are based on general relativity. Quantum physics and general relativity just don't seem to get on.

The problem has led many physicists to investigate some mind-boggling new theories that each give rise to their own version of quantum gravity in the high-energy conditions of the very early Universe. These theories include supersymmetry, which posits links between the different families of subatomic particles. String theory is another possibility. This supposes that everything normally regarded as a subatomic particle is in fact made of tiny vibrating loops, or

1. The weak gravity of our Sun (left) has little impact on the fabric of space. A dense neutron star (centre) makes a bigger dent, while a black hole (right) leaves a chasm in space. 1

strings, of energy. For string theory to work, there have to be either 10 or 26 dimensions of space, depending on the version of the theory that you're working with. Stranger still is M-theory, where the culprits are 'membranes' – like strings, but extended in two dimensions instead of only one, to create quantum-sized 'sheets' of energy from which the more familiar subatomic particles are built up.

At the lower energies of the Universe today – where the quantum side of gravity doesn't show its face – some of these theories boil down to general relativity. However, some lead to more exotic low-energy gravity theories, in which the fundamental 'constant' of nature that governs gravity can vary subtly throughout space and even time. These gravitational variations might be so slight that they could exist without us knowing that they are there.

Voyages of discovery

That could soon change. In 2002, NASA is launching its Gravity Probe B spacecraft – a mission to investigate one of the few remaining untested predictions of general relativity, an effect called 'frame dragging'. This says that massive objects which are rotating should drag space around with them as they go. Gravity Probe B will sit in Earth orbit and look for frame dragging around the planet using a set of four high-precision quartz gyroscopes.

Some physicists think that if gravity does indeed deviate from the predictions of general relativity, then these deviations should show up most strongly in the way that gravity deals with rotation.

1.The Hubble Space Telescope pointing at the spiral galaxy M100 (NGC 4321). Launched into space in 1990, the telescope has revolutionized optical astronomy.

2. Computer simulations of a network of cosmic strings. This is how the network would have looked when the Universe was a tiny fraction of a second old.

So if relativity really is wrong, the key evidence should be locked away in frame dragging, and Gravity Probe B should stand a good chance of routing it out. Establishing the right theory of gravity that holds sway over the Universe today would be a significant step to uncovering the full quantum theory of gravity that ruled the Big Bang.

Over the coming years further space missions, as well as observatories and experiments on the ground, promise to shed light on some of the other outstanding mysteries of modern cosmology. For example, does 90 per cent of the Universe really exist in the form of invisible dark matter? And if so, what is this matter really made of? Will the Universe really expand for ever, or might it still re-collapse in a Big Crunch? What is the correct version of events that led to cosmological inflation? What is the precise form of the irregularities in the density of matter throughout space that inflation produced? And how exactly did the galaxies then grow from these irregularities?

For the astronomers and physicists who spend their days, and nights, piecing together our picture of the cosmos, the most exciting discoveries will probably be the ones that they can't predict – the chance finds. Yet many theorists have plumbed, and are still plumbing, the laws of physics to try to second-guess serendipity. They have uncovered strange possibilities for our Universe – some of which might just turn out to be true.

STRANGE POSSIBILITIES

Some astrophysicists believe that our Universe could be criss-crossed by very long pieces of string. They're not talking about string in the normal sense of the word, but about long, tube-like lengths of high-energy material left over from the Big Bang. These objects are known as cosmic strings. Scientists estimate that the visible Universe could harbour as many as 1000 loops of cosmic string, and could have some 10 long strings stretching right across it. Each string is thought to be about as thick as an atom, but so dense that a single metre of it would weigh as much as the Earth.

Strung out

Strings are thought to have formed during what is called a phase transition in the early Universe, just 10 billionths of a billionth of a billionth of a billionth of a second after the Big Bang. A phase transition represents a wholesale shift in the properties of any substance – be it the stuff of the early Universe or something as simple as water. For example, when gaseous water (steam) cools, it goes through a phase transition as it condenses into liquid water. Keep cooling it and it goes through another phase transition as it becomes solid ice.

As a bucket of water freezes, ice crystals start to form at random locations throughout the water. The crystals grow until they bump into other crystals, and when they all meet up, the whole bucket of water has frozen and the phase transition is complete. The water molecules within any one ice crystal are all aligned in the same direction. But this direction is not the same for every crystal in the bucket. This creates boundaries where the crystals join up, marking regions in which the frozen water molecules are aligned differently.

Anywhere in a bucket of ice where three regions of different alignment meet up, the boundary between them takes a line-like form. Similarly, when the Universe undergoes a phase transition, regions with different 'alignments' can also be divided by line-like boundaries. And it is these line-like boundaries that are known as cosmic strings.

Cosmic phase transitions marked the splitting up of the forces of nature to become distinct entities as the Universe expanded and cooled. In different regions of the Universe the forces split

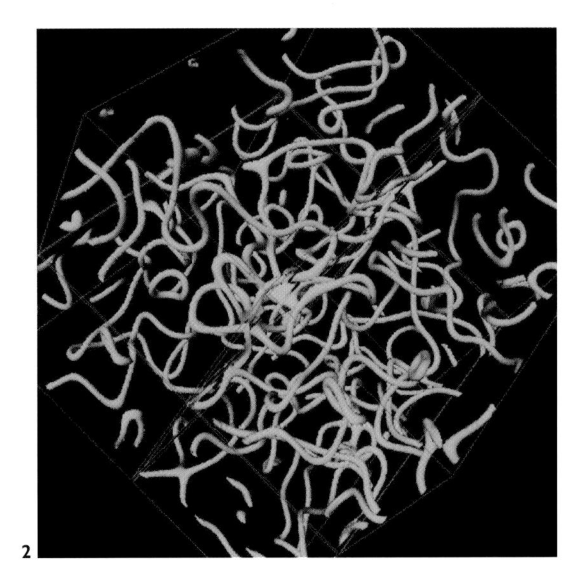
2

apart in slightly different ways. And each region's 'way' can be thought of as its particular alignment.

Strings are just one of several cosmic boundary objects, known as 'topological defects'. Where two regions of different cosmic alignment come together, a 2-D sheet-like topological defect called a domain wall is formed. Four regions meeting up create a point-like defect known as a monopole.

Strings are the most interesting type of defect because some cosmologists have suggested them as an alternative source of the density irregularities, visible in the microwave background radiation, from which galaxies formed. As each string moves through space, its gravity attracts nearby matter, focusing the matter down into great sheets that would trail out behind the string like the wake from a ship. These sheets would then attract more matter by their own gravity, which, so the theories go, would eventually pile up enough to make galaxies.

Most cosmologists now think this scenario is improbable. Studies of the microwave background radiation, by spacecraft such as NASA's COBE probe, have shown that the actual form of the irregularities is inconsistent with those predicted by string-based theories. Most believe instead that the density irregularities that spawned the galaxies were caused by inflation (▷ Chapter 2).

Objects analogous to cosmic strings have been seen forming inside liquid crystals in laboratories on Earth.

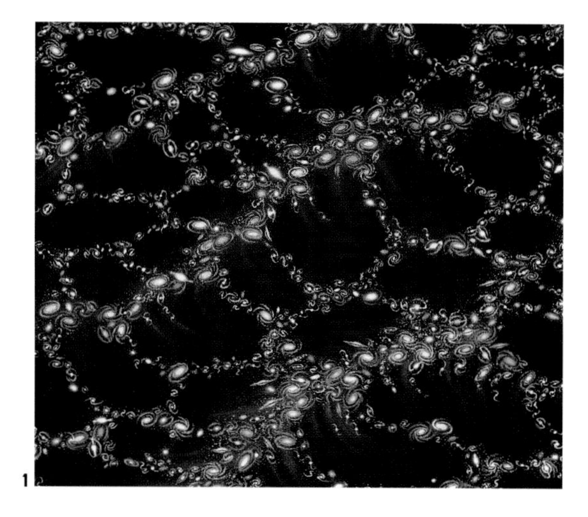

1. Computer graphic of the distribution of galaxies. Sheets and filaments of matter are separated by vast empty voids.

2. Technicians test out magnets for the Large Hadron Collider (LHC) particle accelerator at the Swiss particle physics lab, CERN.

Hyperspace

We are used to living in a Universe that has three dimensions of space. But some theoretical physicists have proposed that there could be extra space dimensions, hidden from our view. This concept of 'hyperspace' has been glamorized in science fiction. But is it grounded in science fact?

Extra dimensions were first incorporated into a serious scientific theory in 1919, by a German physicist, Theodor Kaluza. He realized that Einstein's general theory of relativity and James Clerk Maxwell's theory of electromagnetism could be neatly combined mathematically. There was just one small catch: an extra dimension of space was needed, bringing the total number of dimensions, including time, up to five. In 1926, Kaluza's idea was given a make-over by Swedish physicist Oscar Klein to bring it in line with the new discipline of quantum physics. The result, known as Kaluza-Klein theory, basically says that electromagnetism is down to funny business going on in the fifth dimension.

2

Today, Kaluza-Klein-type theories still exist, but in their modern incarnations as string theory (nothing to do with the theory of cosmic strings) and M-theory. The remit has widened far beyond unifying electromagnetism and gravity, to include both the weak and the strong nuclear forces. And this has raised the number of extra dimensions required from one to, in some cases, almost 30.

The reason we don't see these extra Kaluza-Klein dimensions – if, that is, they really do exist – is that they are rolled up very tightly or, in the language of physics, 'compactified'. A two-dimensional sheet of paper can be compactified by rolling it up tightly and placing it at a distance – do that and it looks one-dimensional. Physicists believe that if extra dimensions exist, they too are rolled up into cylinders, with diameters less than the size of an atomic nucleus. And that is why we can't see them.

At present, there is no direct observational evidence for hyperspace. But the Large Hadron Collider (LHC) – a powerful particle accelerator at CERN in Switzerland – could change that. The LHC, due to be switched on in 2005, should be sensitive enough to detect the echo produced when some types of subatomic particle rattle around inside compactified extra dimensions. Its findings could change the way we think about the fabric of space and time that underpins our Universe.

Mirror, mirror

One startling possibility thrown up by particle physics, notably some versions of string theory, is that our entire Universe might have a twin – a mirror image superimposed on top of itself. This idea emerged in the 1980s. If the theories are to be believed, this looking-glass world is populated with a strange kind of material – 'mirror matter' – that interacts with the material in our Universe only through the action of gravity. The other forces of nature – electromagnetism and the strong and weak nuclear forces – can't penetrate from one side of the mirror to the other. Therefore, light, itself a manifestation of electromagnetism, can't pass through the mirror, so mirror matter is invisible.

In February 1999, scientists exploited this fact to speculate that mirror matter might make up much of the Universe's dark matter. Rabindra Mohapatra and Vigdor Teplitz of the University of Maryland claimed that the haloes of galaxies could be orbited by swarms of stars made from mirror matter, each weighing about half the mass of the Sun.

There is tentative evidence that mirror matter might actually exist. In 1990, physicists at the University of Michigan made atoms of positronium – each consisting of a negatively charged electron orbiting around an anti-electron, also known as a positron. The type of positronium they made is unstable, and should decay after 142 billionths of a second. But the Michigan team found the lifetime of the positronium to be about 0.1 per cent shorter.

In May 2000, Sergei Gninenko of CERN and Robert Foot of the University of Melbourne published a theory that could explain the observation. They calculated that positronium should, in fact, hop back and forth between our Universe and the mirror world. At any time, a number of positronium atoms are off in the mirror world, lowering the number in our Universe – and so making it look as if the atoms are decaying faster. Foot and Gninenko plan further research.

One extra universe probably isn't too much to live with. But some cosmologists are increasingly convinced that our Universe might be just one of many in an ever-multiplying network of parallel universes, which they call the multiverse.

1. Many worlds. An artist's impression of parallel universes, connected by wormholes through hyperspace.

2. Parallel universes. Each universe is shown as a bubble, with stars and galaxies dispersed across the surface.

THE MULTIVERSE

Anyone who has watched *Star Trek* or *Sliders* on TV is already familiar with the idea of parallel universes – other worlds that are almost like our own, except...slightly different. In some parallel universes you might have written this book, and I'll be the one trying to make sense of it all. In others you or I might not even exist. And in others still, life on Earth might never have arisen. But do parallel universes exist in reality? Some physicists think so. The idea was first put on a scientific footing in 1957 by Hugh Everett, a student at Princeton University.

Many worlds

What Everett proposed was a new way of interpreting results thrown up by quantum theory. The laws of the quantum world deal only with probabilities. They can't say with any certainty that

2

'particle x will be here', but only: 'There's a certain probability that particle x might be in this area.'

The odds of a particle being found at any given point in space are determined by what is called its quantum wave function. This is a wobbly 2-D curve that looks like ripples on a pond. The height of the ripples at any particular point in space gives the probability of finding the particle at that point.

When the particle is measured experimentally, its position suddenly becomes well defined, and probability no longer applies. The wave function is then said to 'collapse' – quantum physics essentially switches off, the wave function ripples go flat and are replaced by a single sharp spike at the particle's location. Collapse of the wave function is a central pillar of the traditional interpretation of quantum theory peddled by physicists since the 1920s.

But Everett had other ideas. The crux of his proposal, as applied to the position of a subatomic particle, was that there exist different universes, corresponding to different possible positions that the particle can occupy. Before a measurement is made, these universes are all superimposed. All the possible particle positions then overlap, and that is what forms the particle's wave function. The universes are said to 'interfere'. But when a measurement is made, the interference is disrupted. All the universes split and go their separate ways. Just one universe remains. The person making the measurement then sees the particle as it is in this remaining universe, with a single definite position.

This interference between universes, argued Everett, is what is responsible for the hazy concept of quantum probability. Say, for example, that there are 1000 universes in which the subatomic particle is at position 'a', and only 10 in which it is at position 'b'. When the particle is measured, and the interference breaks apart, we are 100 times more likely to find ourselves in one of the 1000 position 'a' universes, than in one of the 10 universes in which the particle is at position 'b'.

If Everett's idea is correct, our Universe is just one small twig on an eternally branching transuniversal tree that physicists call the multiverse. Each time a quantum event takes place, the multiverse branches into every possible

1

1. Planet Earth is the proverbial drop in the cosmic ocean. But if the multiverse idea is to be believed, our entire Universe is equally insignificant in the scheme of things.

2. Some cosmologists suggest that the Universe could be flat, or even spherical; others put forward the bizarre possibility that it might be doughnut-shaped.

2

outcome of the event. Our own Universe follows just one branch. Think of all the atoms and other quantum particles on Earth, in the Solar System, in our galaxy and in the Universe. Think how many quantum events must take place every second. Now recall that a billion billion seconds have elapsed since the Big Bang. The multiverse is truly vast.

Everett's so-called 'many worlds' interpretation of quantum theory is appealing because it dodges philosophical glitches, such as why, in the collapsing wave function view, quantum physics suddenly switches off when a measurement is made. In many worlds, it doesn't switch off: the act of measuring simply disrupts the interference between universes that causes quantum effects.

Evidence for the multiverse theory has emerged from computing. Scientists are building quantum computers – devices that harness the computing power of their counterparts in parallel universes.

SPACE FOR HUMANITY?

Some physicists have suggested that the many worlds view could provide a basis for the quantum theory of gravity that underlies the Big Bang. Others say it could offer a way out of the paradoxes that have plagued theoretical attempts to devise time machines – paradoxes such as travelling into the past and killing one of your parents before you were born. Proponents of many worlds say that you would go back into the past of an alternative universe, so your actions would have no effect on the universe that you had left behind.

Some researchers have even supposed that the parallel universes in many worlds quantum physics could help explain our very existence.

The anthropic Universe

In 1957, Princeton physicist Robert Dicke published a scientific paper in which he placed constraints on the size of our Universe, based on biology. He showed how the Universe had to be of a certain size if life were to have evolved within it.

The idea that whatever goes on in the Universe must – by the fact of our existence – be consistent with the emergence of life, has become known as the anthropic principle. It is a powerful deductive tool. Also in 1957, a team led by British astronomer Sir Fred Hoyle used the principle to predict correctly the existence of a hitherto unknown nuclear reaction. When this reaction occurs inside stars, it produces carbon. Without it the huge amounts of carbon needed for life on Earth cannot be produced.

In his book *Just Six Numbers* (1999), Britain's Astronomer Royal, Professor Sir Martin Rees of Cambridge University, takes the idea further. He argues that biological factors so constrain the properties of the entire Universe that it appears to be 'fine-tuned' so that life can emerge within it.

Rees believes that the large-scale Universe is defined by six key parameters (numbers), which are actually denoted by letters: D, the number of space dimensions in the Universe that aren't compactified (▷ pp. 86–7); N, the relative strengths of electromagnetism and gravity; E, the energy

1

released by nuclear reactions; Q, the size of the irregularities in the microwave background from which the galaxies grew; Greek letter lambda, λ, which represents the amount of dark energy (▷ pp. 66–7) pervading space; and lastly omega, Ω, the cosmic density parameter. Rees argues that if any of these parameters differed greatly from their observed values, we wouldn't exist.

The human condition

Rees believes that cosmic fine-tuning can be explained naturally within the framework of the multiverse. There isn't only one Universe – in which case fine-tuning would be a big coincidence – there are many. Each universe takes different values of the six magic parameters, so somewhere there will be a universe with a set of parameters just right for our kind of life. It's no coincidence that this is where we find ourselves – we couldn't exist anywhere else.

The bottom line might seem a bit bleak. There is nothing inherently special about life on Earth, or our place in the cosmos. We weren't put here for any purpose. We simply emerged from the right blend of chemical goo in a corner of the multiverse where the conditions were, by chance, hospitable.

But as species go, we haven't done so badly. We have unravelled the story of our own beginnings – the Big Bang theory. And that's something to be proud of. As the physicist and author Stephen Hawking once put it: 'We are just an advanced breed of monkeys on a minor planet of a very average star. But we can understand the Universe.

'That makes us something very special.'

2

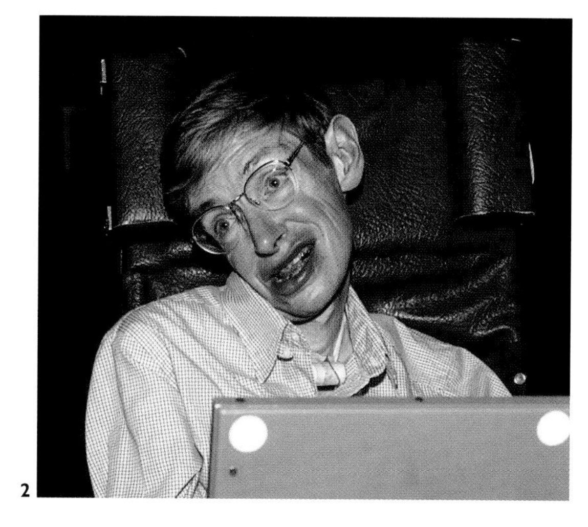

1. Professor Sir Martin Rees of Cambridge Univesity has investigated whether ours is one of a multitude of 'universes'.

2. Professor Stephen Hawking has greatly contributed to the understanding of our Universe and beyond.

Physicist Richard Feynman once likened understanding the Universe to learning chess – you can soon pick up the rules, but you're still a long way from becoming a grandmaster.

FURTHER INFORMATION

Books

John Gribbin, *In Search of the Big Bang* (Penguin Books, 1998). An accessible account of today's understanding of the Universe.

Carl Sagan, *Cosmos* (Warner Books, 1983). Still one of the most enthralling popular books on space and the Universe.

Stephen Hawking, *A Brief History of Time* (Transworld Publishers, 1988). A popular science classic, written by one of the world's greatest cosmologists.

Robert Wilson, *Astronomy Through the Ages* (Taylor & Francis, 1997). A history of how humans have observed the Universe from the end of the last ice age to the present day.

David Deutsch, *The Fabric of Reality* (Allen Lane, The Penguin Press, 1997). An intriguing view of the Universe and what lies beyond.

Cliff Pickover, *Surfing Through Hyperspace* (Oxford University Press, 1999). One of the most readable accounts of how our Universe may harbour extra, unseen dimensions of space.

Lee Smolin, *Three Roads to Quantum Gravity* (Weidenfeld & Nicolson, 2000). In this popular book, physicist Lee Smolin reviews current attempts to unify the laws of the quantum world with gravity.

Martin Rees, *Just Six Numbers* (Weidenfeld & Nicolson, 1999). Britain's Astronomer Royal, Professor Sir Martin Rees, suggests that the Universe can be described by a handful of numbers.

Monthly Magazines

Astronomy Now (Pole Star Publications Ltd, UK).

Astronomy (Kalmbach Publishing Co, USA).

Websites

Astronomy Now
http://www.astronomynow.com

SpaceRef
http://www.spaceref.com

SPACE.com
http://www.space.com

Hubble Space Telescope public information and pictures
http://oposite.stsci.edu/pubinfo

Big Bang research at Cambridge University
http://www.damtp.cam.ac.uk/user/gr/public/bb_home.html

Europe's Planck cosmology spacecraft
http://astro.estec.esa.nl/SA-general/Projects/Planck

FAQ on cosmology by NASA
http://map.gsfc.nasa.gov/html/web_site.html

Guide to telescopes around the world
http://www.vilspa.esa.es/astroweb/yp_telescope.html

INDEX

Italic type denotes illustrations